Light: A Very Short Introduction

Very Short Introductions available now:

Available soon:

For more information visit our website

www.oup.com/vsi/

Ian Walmsley

LIGHT

A Very Short Introduction

OXFORD

UNIVERSITY PRESS

Great Clarendon Street, Oxford, OX2 6DP,
United Kingdom

Oxford University Press is a department of the University of Oxford.
It furthers the University's objective of excellence in research, scholarship,
and education by publishing worldwide. Oxford is a registered trade mark of
Oxford University Press in the UK and in certain other countries

© Ian Walmsley 2015

The moral rights of the author have been asserted

First edition published in 2015

Published in the United States of America by Oxford University Press
198 Madison Avenue, New York, NY 10016, United States of America

British Library Cataloguing in Publication Data
Data available

Library of Congress Control Number: 2015937261

ISBN 978–0–19–968269–0

Printed and bound by CPI Group (UK) Ltd, Croydon, CR0 4YY

To Kate, Alex, Nathaniel, and Clair

Contents

Preface and acknowledgements

Light enables us to see things. This tautology is the starting point of a long and fascinating trail from which developed both extraordinary views of how the natural world behaves and the very tools and methods on which modern science is built—explanations that are commensurate with experiments.

Light carries information about our surroundings, from distant stars and galaxies to the cells in our bodies to individual atoms and molecules. It is the basis of many technologies that enhance our quality of life: the Internet is powered by light; the most precise clocks in the world rely on light; the tiniest objects, from individual atoms, to live biological cells, can be observed and manipulated using light; images and displays are everywhere. Light reveals the full strangeness of the quantum world and inspires our imagination about the world. Yet it might surprise you to know that what light actually is was only really understood less than a hundred years ago, and even now we are eking out new insights from that understanding.

This book explores how people have come to our current view of what light is and what it does. It's a great story, reaching back into the ancient world, with a global cast of contributors, from Euclid in Athens and Al-Hazen in Baghdad, conceiving the idea of light rays, to Ted Maiman in Los Angeles and Shuji Nakamura in Tokushima developing new lasers, by way of Joseph Fresnel in Paris and Thomas Young in London thinking about light waves, to James Clerk Maxwell in Aberdeen and Heinrich Hertz in Berlin developing from these ideas the concept of electromagnetic fields, that finally culminated in Albert Einstein in Bern and Paul Dirac in Cambridge explaining these apparently mutually incompatible facts by a radical new concept—the quantum field—that showed how it is possible to be both a particle and a wave at the same time.

At each step, new understanding (for example, of refraction) has led to new applications (such as eyeglasses for correcting vision): the path from discovery to technology is often very short when light is involved. The impact of light on the modern world is immense, and often unrealized. For this reason, 2015 has been designated by the United Nations as the International Year of Light: a celebration of light and what is made possible by it.

I should like to record my thanks to Alex Walmsley and Latha Menon for helpful comments on drafts of this book, and to many colleagues for responding to specific technical questions as well as those who kindly provided illustrations for the figures. Errors and statements that go beyond the exhortation, attributed to Einstein, to make things as simple as possible, but not simpler, are due to me.

INTERNATIONAL
YEAR OF LIGHT
2015

List of illustrations

Light

Chapter 1
What is light?

Light enables us to see the world around us. It provides the means by which our sense of sight gleans the most direct information about the physical arrangement of the world and how it is changing. Indeed, the capacity of light to carry and convey information is perhaps its most important, and remarkable, characteristic.

Seeing is believing

Sight enables us to locate ourselves in our surroundings, defining things outside ourselves that allow us to construct a true picture of the world. And sight inspires the imagination beyond the physical sensation of vision itself. George Richmond's painting, *The Creation of Light*, shown in Figure 1, illustrates the central place light has in our psyche. Indeed, words deriving from the idea of light—insight, illumination, clarity, for example—pertain to human, as well as physical, qualities. In fact, Latin has two words to describe light, *lux* and *lumen*, denoting both the material and the metaphysical aspects of light. It is the former with which this book is primarily concerned. The intertwining of the physical and the poetic has made light a metaphor for thinking about the world, in philosophy, theology, psychology, art, and literature. Because it is something of which almost everyone has direct experience, the physical basis of light and how it

1. *The Creation of Light* by George Richmond.

facilitates this powerful sense has made it an object of study for philosophers and more recently scientists, for centuries.

Light gives life. Literally, light plays a vital role in the biological and chemical processes that underpin our and our planet's existence. Figuratively, light frames our perception of our surroundings. Our common everyday experiences illustrate the central importance of light in this regard. Of course we use it for illuminating our environment, either naturally, by the Sun or Moon, or artificially. Most common light sources use electricity, but on occasion we still use chemical reactions to generate light; burning candles, for example. The different character of the illumination has an effect on how we perceive our surroundings: it sets a 'mood' for the physical space.

Light has a very fundamental role in making possible life itself. The primary source of energy for the Earth is ultimately the Sun. And the means by which energy is transmitted from the Sun is light, both the visible components we can see, as well as some invisible ones we cannot see directly. Think, for example, of lying on a beach or sitting out in the garden in the sunshine: the warmth we feel is a consequence of some of the 'invisible light' radiating from the Sun. This is just one example of the physiological effects of light.

But at the very core of our planet's ability to sustain life is an awesome biochemical process that converts 'waste' molecules— carbon dioxide—into 'useful' ones—oxygen, using sunlight as a source of energy. The reverse process, conversion of oxygen into carbon dioxide, occurs during respiration, as well as from the things we burn to power our everyday world.

The Sun's light has over the course of millions of years enabled the formation of the current biosphere and the geology that provides other sources of energy. Neither coal nor oil could have been made without energy from the Sun. And our use of these resources is changing the way in which light from the Sun affects our planet. Some of the invisible light from the Sun—ultraviolet light—is still absorbed by the Earth and its atmosphere. But the other invisible part—infrared light—is reflected back by atmospheric gases. By the same mechanism, infrared radiation is trapped on the planet, contributing to increased planetary surface temperature.

Light enables communication

Pictures have been part of human culture since the beginning of the species. The impact of images on how we conceive of the world, and how we make sense of our place in it, is immeasurable. Optical technology has contributed to this in ways that are utterly transformative. For instance the ability to capture images easily and rapidly, by means of film-based and digital photography, allows us to record places, people, and things as reports that can be widely distributed (nowadays by an optically enabled Internet) and which have lasting impact: images of leaders and workers, awe-inspiring scenes of the natural world, and horrifying scenes of war. These can bind or fragment people in unexpected ways: calling populations to action, inciting acts of compassion, and giving deeper insight into shared experience. Recall the astonishing sense of wonder the sight of man's first steps on the Moon (see Figure 2) invoked. The ability to capture moving images adds a completely new dimension by enabling new narrative and documentary capabilities. Can you imagine life without television or movies, without video?

Nowadays, the generation and transmission of images is so prevalent we hardly think of it. We use self-luminous displays every day: televisions, computers, tablets, even smartphones. All of these bring information to you and receive it from you using light as a medium. Almost all long-distance telecommunications travel on light beams, guided along thin strands of glass called optical fibres. This is the basis of fibre optical broadband services that link our homes to the Internet. Even inside computers and televisions light plays a role. For instance, the music, video, or images that are locked in a CD or DVD are accessed using light. A tiny moving head based on a miniature laser 'reads' the disc and converts the information coded on it into electrical signals that can then be sent to the display screen. All our surfing, downloading, and emailing activity now requires such immense information capacity that light is the only feasible medium for conveying it.

2. Neil Armstrong's photograph of Edwin Aldrin walking on the Moon.

Transportation in the modern world uses light as the means by
which we signal and regulate our movements. From streetlights in
towns, to landing lights on aircraft, light is an essential part of
navigation. And it even plays a role in maintaining our vehicles.
For instance, lasers are used to align the wheels of a car, and
light-enabled distribution of ignition power to drive an internal
combustion engine is not uncommon.

In many, many ways, light carries the energy and information that makes modern life possible.

Optics

The field of enquiry that constitutes the study of light is called optics. Optics is among the oldest of the sciences, and its historical development forms one of the most important paths in the emergence of modern science. Ideas arising in optics have stimulated new ways of thinking for very disparate fields, such as the mechanics of motion of atoms and molecules. And technologies enabled by deeper understanding of light have been central in unlocking other secrets of the natural world. Galileo's telescope designs, for instance, were critical in his observation of the moons of Jupiter, which was a vital step in moving towards a view of the solar system in which planets moved around the Sun. This, in turn, was important in developing the laws of gravity that govern planetary motion.

The origins of optics lie in the work of the Greek philosophers of the 4th century BCE, and the field has continued to flourish for the past two millennia. It is perhaps surprising that we can still discover new things about light after such a long period of attention by many clever people. Yet optics remains at the forefront of current science: more than ten Nobel Prizes in the past twenty years have been awarded for research in which light has played a central role, from controlling and measuring the motion of atoms and molecules, at unimaginably low temperatures and on timescales of breathtaking brevity, to improving the precision of clocks a thousandfold, to enabling us to look inside living cells and watch what happens as they change.

What is light?

A good place to start is with some of the things commonly associated with light: brightness, intensity, colour, and warmth.

These are all tangible properties that suggest light is a physical entity. But what, exactly, is it?

We may consider what it means for light to be bright by taking a particular source of light—a household light bulb. These come in various sizes, but all have *powers* of several tens of Watts (the unit in which power is measured, labelled W, and signifying the energy consumed per second of operation). A 50 W light bulb gives sufficient light by which to see things inside a house. Car headlights are typically of slightly greater power, approximately between 60 W and 100 W. The floodlights at a football ground are of much larger power, up to several thousand Watts. I will discuss later exactly how these different sources generate light, but these powers will give a sense of the brightness of the corresponding lighting. Of course, one of the brightest sources is the Sun. It has a massive power output, more than 10^{25} W (1 with 25 zeroes after it), which makes it impossible to look at directly even though it is a very great distance away.

This brings us to the next concept associated with how bright a light is. Light of the kind discussed above looks dimmer the more distant it is. So power alone is not the only criterion determining brightness. It is related in some way to the fraction of the power that we can receive from the light source. For example, a laser pointer typically has much, much lower power output than a light bulb, often only a few thousandths of a Watt (10^{-2} W or 10 mW). Yet it appears very bright when it is pointed at a screen.

What is important here is the *intensity* of the light generated by the source—the power per unit area of the receiver. (This is more properly the *irradiance*, but intensity is perhaps a more familiar term.) The intensity of a light source is related to the ability to concentrate the light. A laser pointer appears to be very bright because its light beam is concentrated on a small spot on the screen, whereas the Sun's light is diffused over a very wide area.

Therefore, although the Sun has a very great power output, the light it produces is not as intense as that of a laser pointer.

The underlying property that describes our ability to concentrate light is called the *spatial coherence*. This is related to the source's propensity to send light in a particular direction. For instance, both the Sun and a light bulb radiate light into all directions—the Sun can be seen from every place on Earth, and a light bulb from wherever you stand in a room. But a laser pointer only puts out light in a single direction—that in which you point it. So you cannot see the laser beam unless you look at the surface on which it is incident. Because of the property that its beam has a well-defined direction, the laser pointer is said to be a coherent light source, whereas the light bulb is an incoherent light source.

Another defining characteristic of light, perhaps its most evident property, is *colour*. The rainbow embodies the fundamental idea of a spectrum of colours—a palette emerging from the conjunction of rain and sunlight—from blue at one end to red at the other. A central strand in the development of theories of light has been the development of models of colour vision. Colour is intimately tied to perception, as well as to physics. An illustration of this is found in the experiments on the nature of colours undertaken by Sir Isaac Newton (Figure 3), a dominant figure of early 18th-century science, and whose book *Opticks* defined the view of light for two centuries, and by Johann Wolfgang von Goethe (Figure 3), a dominant figure of late 18th-century literature, who incorporated scientific ideas in his writings, but nonetheless believed Newton to be profoundly wrong about the nature of light.

The first part of Newton's famous experiment (similar to ones undertaken by Descartes and others previously to him) was to allow a small beam of light from the Sun, defined by a hole in a dark screen, to pass through a prism, and thence to fall upon a screen. The familiar rainbow colours emerge. Goethe was fascinated by this effect, and borrowed some prisms from a local

3. Isaac Newton (left), Johann Wolfgang von Goethe (middle), and Rosalind Franklin (right).

aristocrat with which to experiment for himself. He concluded quickly that Newton's experiment was wrong—in the sense that his claim of the universality of colour dissection from white light was not true. Goethe had himself discovered a very different set of colours.

The experiment Goethe undertook was to look through the prism at the mullion of a window. That is, he looked at a dark line against a bright background, the very opposite of what Newton had done. What he saw was a very different spectrum from Newton. Not the red, green, and blue of the Newtonian spectrum but rather a new palette of cyan, magenta, and yellow; the so-called complementary colours of the Newtonian spectrum. Combining Newton's colours produces a white image—combining Goethe's produces a black one.

Goethe espoused the idea that colours are the things perceived; Newton defined them as intrinsic properties of light. They were both right. Nowadays we are content to separate the physical attribute of colour from its physiological effects, the sensation of colour. We each react differently to colours—indeed, coloured light can even be used as a therapy. From an artistic point of view, the interpretation that our consciousness realizes from the sensations associated with light of a particular colour is a critical matter—perception is vitally important. Yet from a physical point of view there is an underlying property that we can assign to the label 'colour' unambiguously: its frequency—at least until we get into the realm of quantum light.

The reach of light extends beyond the visible spectrum, at the blue end into the invisible realm of the ultraviolet and the extreme ultraviolet to X-rays and γ-rays. At the other lie the infrared, microwaves, radio waves, and eventually T-rays (Figure 4). To see these, we need different instruments than our eyes alone. Nonetheless, we know that light of these colours exists. For instance, the warmth of the Sun is due to infrared light that is absorbed by our skin. At lower frequencies microwaves are used for cellular phone communications as well as for cooking, by heating the water in the food being prepared. The invisible colours at shorter wavelengths are also familiar. Sunburn is caused by ultraviolet light, while X-rays are used routinely in medical imaging.

X-rays are also used in many non-medical applications. For example, the patterns made when X-rays scatter from a regular arrangement of atoms in a molecule or solid enable that arrangement to be determined, even though the atoms are spaced very, very close together, at distances more than 10,000 times smaller than a human hair. X-ray diffraction images can reveal the very structure of the molecule or solid, with profound implications. Perhaps the most famous example is the structure of DNA molecules identified by James Watson and Francis Crick more than half a century ago, based on X-ray images taken by Rosalind Franklin (Figure 3) and Maurice Wilkins. Knowing how molecules replicate revolutionized biomedicine.

These applications are indicative of the importance of light—in its broadest sense—in making possible our modern world, and impact our ability to enjoy it to its full. They rest on the fundamental work of a number of scientists in the 19th century—Michael Faraday,

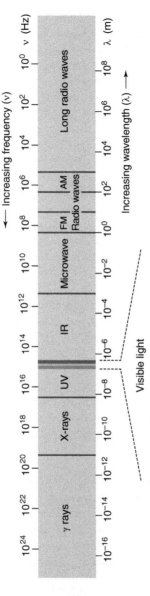

4. The spectrum of electromagnetic waves.

Hans Christian Oersted, André-Marie Ampère, Charles Augustin de Coulomb, Alessandro Volta, Georg Ohm, James Clerk Maxwell, and Heinrich Hertz. That there exists a connection between visible light and other apparently disjoint things like microwaves and X-rays is remarkable, and it was the triumph of scientific enquiry by these men and others who made and identified these connections.

The gamut of colours, or spectrum, provides a tool for art and science. Whereas a painter or artist explores ways in which colours themselves are juxtaposed or combined, a spectroscopist explores ways in which matter responds to different colours. For example, in the early 19th century Joseph von Fraunhofer determined some of the types of atoms that are present in the Sun by looking carefully at the particular colours of light that the Sun emits. He noticed that characteristic colours were missing from the Sun's spectrum and noted that these colours were 'fingerprints' for particular atoms. The study of spectra is the domain of spectroscopy, which uses light to identify different atoms or molecules. It is a vital research activity today, with impact in many areas from health monitoring to remote sensing of the atmosphere for pollutants.

Apart from these familiar properties of light, there is one further property I want to point out. It, too, is something that we all know about from our everyday lives, though perhaps in a less explicit way than we experience the other properties of light. It is polarization.

If you have watched a 3D movie, then you have seen this property being exploited. Watching such a movie requires you to wear special spectacles with cardboard or plastic frames that have pieces of plastic for the 'lenses'. If you take two pairs of these spectacles and slide the left lens of one over the right lens of the other, then look through them at a light bulb, you will see a very,

very dim image of the bulb. Alternatively if you rotate one of the spectacles through 90 degrees with respect to the other and put either the two left lenses or the two right lenses over one another you will see something very similar—almost no transmission of light through the pair of glasses.

This can be explained by assigning a property of 'orientation' to the light. Most common light sources emit light without a preferred orientation. When you look at a light through these spectacles it will appear dimmer, indicating that they have selected a particular orientation. The left lens allows one orientation to pass, the right lens an orientation that is at right angles to the first. That's why when you put a second lens oriented at right angles to the first, nothing will be transmitted through it: the light passing through the first lens has the 'wrong' orientation to pass through the second. This characteristic of orientation is called polarization. It took a great deal of careful exploration to come up with and to understand the idea of polarization, but it is an important feature for technologies based on light, and indeed for understanding what light actually is.

These physical characteristics of intensity, colour, and polarization are what enables light to be used to discern, to measure, and to control properties of material substances, and therefore to form the basis of a host of tools for studying and manipulating matter and even small objects. In the examples above, light almost always plays the role of an information carrier. Whether it is conveying an image or a spectrum or a phone conversation, it acts as a messenger. But light can also be used in an active way. For instance, the heating properties of light can be used to cut metal and other materials very precisely. Thick plates of metal, up to a centimetre or so, can be machined using high-power lasers more rapidly, with a higher quality finish and less waste, than using a saw. And light is used in medicine in various ways, from correcting vision through laser surgery to activating drugs for anti-cancer therapies.

Light enables us to view the natural world at every time scale and distance scale imaginable; from the earliest moments of the universe to the unimaginably fast motion of electrons in atoms and molecules; from the large-scale arrangement of clusters of galaxies across the universe to the atomic arrangement of carbon atoms in graphene. It provides us with insight into the very foundations of the natural world, from the weirdness of quantum physics to the structure of DNA molecules.

The story of optics is one in which new discoveries about light have enabled new technologies that have, in turn, given rise to new discoveries across many fields of science. At each stage, from the invention of eyeglasses to the most precise atomic clocks, to modern technologies for imaging, measuring, and communications, light has given new applications that have revolutionized how we live. This cycle of discovery and innovation makes the study of light a vibrant discipline, even as it is among the most venerable. How we came to a modern understanding of what light is, and therefore how we could use it, both for new understanding of the world and for new capabilities that change the world, is the tale that forms the rest of the book.

Chapter 2
Light rays

Taking a selfie requires that the phone camera is pointed towards you. It's obvious that this is necessary if you are to be in the picture. But this simple fact indicates something about the nature of light: to see an image of an object, there has to be a straight line between the object (in this case you) and the camera lens. This is usually called the 'line of sight'. Thus light is something that propagates in a straight line from the object to the viewer.

Indeed, this is what we might expect from our knowledge of certain types of light source. Eye-catching visual effects at concerts are generated using lasers, illuminating the stage and the performers with coloured light beams. Laser pointers are commonly used at talks or lectures to emphasize images or words on a screen. The beam produced by these coherent light sources is highly focused, hardly diverging at all even across a big hall. It goes in a straight line—you point the device in the direction where you want the light to go, and it does so.

Because sunlight does not obviously exhibit this characteristic, it required some thinking to determine that the propagation of light in a straight line was exactly what was needed to understand why distant objects appeared to be smaller than nearer ones, even when it was known that they were, in fact, exactly the same size physically.

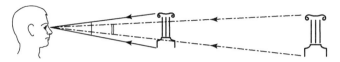

5. Euclid's construction of rays showing why objects of the same size look smaller when they are further away.

The insight that the concept of straight-line propagation could explain this effect is attributed to Euclid, working in Greece in around 300 BCE. His idea, from one of the earliest books on optics, is illustrated in Figure 5. Imagine two lines—let's call them rays—one connecting the top of the object (a pillar in this case) to the observer's eye, and one connecting the bottom of the object to the eye. The angle between these is related to the apparent size of the image of this object that we perceive. A more distant pillar, the same physical size as the first, produces two rays whose angle of intersection at the observer is smaller—hence the pillar appears smaller. This is what we call perspective in the image.

What is the stuff that traverses the rays? Euclid (building on earlier ideas) thought that it was particles sent out from the eye itself (from an imaginary internal fire) that illuminate the object and are reflected back to the observer. But this suggests that we would be able to see things whether or not it was dark outside. Nonetheless, the idea of stuff moving along a trajectory between object and observer remained a powerful concept.

Euclid's idea was modified by Alhazen in the 11th century to a form that we now use routinely: objects are illuminated by rays from the Sun (the external fire) which are scattered towards the observer. There are several stories about how he came to this idea—including doing an experiment when he looked directly at the Sun, and determined that the painful sensation he experienced would be there all the time if the 'internal fire' were burning all the time. Thus, he argued, the source of the light necessary to generate the image was external.

Let us, for the sake of this argument, assume that what moves along these rays are particles of light: call them *photons*. The brightness of the beam is related to the number of photons traversing the ray in one second. In order to understand how an image of an object is formed, we'll need to consider what happens when one of these photons is reflected off a mirrored surface, as well as what happens at a lens. This will lead to the 'laws of optics' that are used for designing very complex optical instruments such as surgical microscopes, and catheters for 'keyhole' surgery, as well as massive optical telescopes placed in orbit above the Earth for observing distant galaxies. The impact of these instruments on our life and our understanding of the world is immense.

What sort of properties do these particles of light possess? The usual sorts of attributes assigned to a particle are its position, its direction of travel, and its speed. For now, assume that it moves at the 'speed of light', without going into detail about what that actually is. The position of a photon might then specify the starting position of the 'ray' and the direction of the photon's motion would be the direction of the ray. The photon heads off from its starting point in this direction at the speed of light until it encounters the surface of the object.

Reflection

When it hits the object, the light is reflected. What happens is that the photon 'bounces' off the surface, changing its direction, but not its position on the surface. The situation is shown in Figure 6. The manner in which the direction is changed is specified by the 'law of reflection', discovered by Hero of Alexandria in the 1st century. It states that the angle of incidence (that is, the angle between the incoming ray direction and the direction perpendicular to the surface at the point of incidence) is equal to the angle of reflection (that is, the angle between the outgoing ray direction and the direction perpendicular to the surface at the point of incidence). There are some surprising consequences of this conceptually simple and yet extremely powerful law.

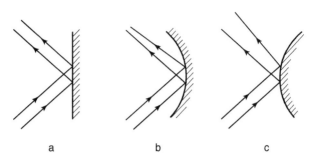

6. Rays of light reflecting off a. a flat mirror and b. and c. a curved mirror.

The apparent left–right inversion associated with reflections follows from this idea. If you hold up your watch to a mirror and look at its reflection, you'll see that its second hand is moving in an anticlockwise direction, and that when you move your right hand, your doppelgänger moves his or her left hand. This change of 'handedness' is the hallmark of the image through the looking glass—the world reflected in a mirror is in that sense really topsy-turvy.

This can be explained entirely by Hero's law of reflection. Figure 7 shows how a mirror generates a reflection with opposite handedness. The clockwise-pointing arrow is the object. Rays from each point on the arrow reflect off the mirror and are rearranged so that the arrow seen in the mirror is pointing anticlockwise. You can use the same construction to show that a left-pointing arrow is seen in the mirror as a right-pointing arrow and vice versa, but an up-pointing arrow remains pointing up, and a down-pointing arrow points down in the reflection.

Imaging using reflections

If, instead of looking at your opposite-handed self in a flat bathroom mirror, you see your reflection in a polished spoon, then you see a

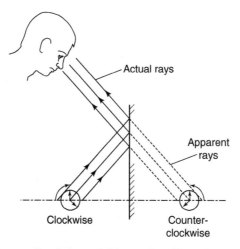

Actual rays

Apparent rays

Clockwise

Counter-clockwise

7. **The change of handedness of objects reflected in a mirror.**

distorted form of your image: magnified features on curved backgrounds. The concave front surface of the spoon magnifies the object and the convex back surface demagnifies it.

Why is this? Since the formation of images is perhaps the most important application of optical instruments—ranging from contact lenses for vision correction to space telescopes for scientific discovery—it's worth understanding exactly how this happens.

Up to now, I've considered only a single ray of light from one point on the object. In fact, rays usually scatter in all directions from each object point. Consider a 'bundle' of such rays all coming from a single point on the object, forming a cone around the original ray. This cone of rays diverges as it moves away from the object, as shown in Figure 8. The rays hit the curved mirror surface at different points and therefore also at different angles of incidence. Thus they reflect in different directions, each one still satisfying Hero's law of reflection.

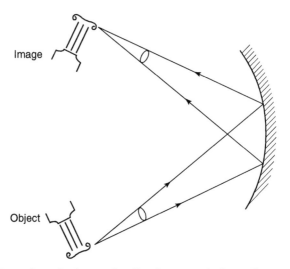

8. Image formation by a ray bundle using a curved mirror. The rays from one point on the object meet at one point in the image.

In fact they now form a converging cone, and eventually all meet at a single point. This is the 'image' point of the original object point.

An image as we normally consider it is made up of multiple such image points arising from different object points. The size of the image is determined by the distance of the object from the mirror and the focusing power of the mirror, specified by the radius of curvature of surface (e.g. a more concave mirror has a smaller radius of surface curvature). The image can be bigger than the object when the object is closer to the mirror than the image. The ratio of image to object size is called the magnification.

The image-magnifying feature of a curved mirror was used by Newton to design a telescope, shown in Figure 9. His design has a remarkable property: it forms images of distant objects that are

9. Newton's reflecting telescope enabled images without chromatic aberration.

the same size for every colour (said to be free from 'chromatic aberration'). Newton cleverly used the property that the angle of a reflected ray for a fixed angle of the incident ray is the same no matter what the colour of the light. The image of each colour is therefore formed in the same place—all the colours register perfectly, guaranteed by physics.

Refraction

Newton invented this instrument because the telescopes used by other early pioneers such as Galileo Galilei and Johannes Kepler suffered seriously from chromatic aberration. The images formed by their telescopes always had a blurred coloured halo around the edge of the object. The reason for this is that they were designed using a different property of light rays—refraction: the phenomenon that light rays bend when they go from one transparent medium to another.

It is the refraction of light that produces the 'kink' observed in a pencil partially immersed in a bowl of water. This is described by the law of refraction, commonly known as Snell's law, after Dutchman Willebrord Snell, an early 17th-century proponent. This law says that the angle the exiting ray makes with a line perpendicular to the surface is related to the angle of the incoming ray by the properties of the two media that form the interface—in our example the surface of the water. This is illustrated in Figure 10.

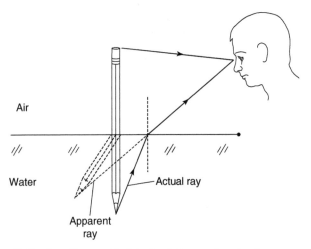

10. Refraction of a ray at the interface between air and water.

The particular property of the media that is relevant here is the 'refractive index'. The refractive index can be thought of as being a measure of the optical 'sluggishness' of the medium as experienced by a light ray. So light travels more slowly in a medium with a larger refractive index because the molecules of the medium are slightly more resistant to having their atoms and electrons moved by the light. It is like running in a pool of water. If the depth is very small, your legs can move easily and you can run fast. If the water is up to your knees, it's harder, because you have to work against the resistance of the water.

In fact, the law of refraction can be derived from this analogy. Pierre de Fermat showed that when light goes from a point in one medium to a point in another medium, it seeks a ray that traverses a shorter time in the medium with high refractive index and a longer time in the medium with a lower refractive index. This requires the light ray to bend at this interface between the two media, and Fermat's principle turns out to be entirely the same as Snell's law.

Imaging using lenses

Now, just as a curved reflecting surface can form an image of an object, so can a curved transparent surface. How this happens is shown in Figure 11. A bundle of rays from a point on the object are brought to focus at the image. Notice the shape that does this—it has the same cross section as a lentil. This is the origin of the word 'lens'.

Lenses are ubiquitous in image-forming devices, from your eyes to mobile phone cameras to surgical microscopes. Imaging instruments have two components: the lens itself, and a light detector, which converts the light into, typically, an electrical signal. In the case of your eyes, this is the retina, whereas for the mobile phone it is an array of minute pieces of silicon that form solid-state light sensors.

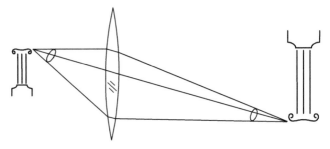

11. Image formation by a ray bundle using a lens.

The lenses in each of these devices are different, of course, but the basic principle is the same for each. In every case the location of the lens with respect to the detector is a key design parameter, as is the *focal length* of the lens which quantifies its 'ray-bending' power. The focal length is set by the curvature of the surfaces of the lens and its thickness. More strongly curved surfaces and thicker materials are used to make lenses with short focal lengths, and these are used usually in instruments where a high magnification is needed, such as a microscope.

Because the refractive index of the lens material usually depends on the colour of light, rays of different colours are bent by different amounts at the surface, leading to a focus for each colour occurring in a different position. This gives an image that has 'haloes' of different colour around it. For example, only one colour may be properly in focus at a particular detector plane; the others will be out of focus, and form the halo. Whether this chromatic aberration is an important effect or not depends on the particular application.

One of the most familiar, and indeed most important image-forming instruments of this type is the eye. It consists of a front-refracting surface—the cornea—and an adjustable lens, which changes shape as you focus on things at different distances. These elements form images on the retina at the back of the eye.

Historically, the formation of images by the eye was of great interest, since an experiment by Descartes (see Figure 12) showed that the image of an upright object was upside down. Of course, we don't perceive the object in this way, so it was clear that the brain undertakes some remarkable processing between the raw retinal signals and the perception of the external world.

Optical instruments

As many of us are all too aware, the ability of the eye to form high-quality images (sharp, undistorted, and in colour) can degrade as we age. One of the earliest applications of optical instruments was developed as an aid to sight under such circumstances. Eyeglasses were perhaps the first optical technology, purportedly invented by Roger Bacon—the 'mad friar' of Oxford—in the 13th century.

The corrective elements are often simple lenses, placed either in a frame at some distance (a few millimetres) from the cornea (the front surface of the eyeball), or 'contact lenses' placed, as the name suggests, in contact with the cornea. In both cases the imaging system is compound, that is, it consists of several elements—the external lens, the cornea and the internal ocular lens. This functional form provides the necessary degrees of freedom to correct most kinds of vision by enabling the external lens to compensate for the imperfections of the internal lens. This can also be done by directly altering the shape of the front surface of the eye by laser surgery. One approach, laser-assisted in situ keratomileusis (LASIK), uses the laser to ablate part of the cornea. This changes its curvature, thereby altering its focusing power, and thus the image-forming capabilities of the eye.

Many other image-forming instruments work on very similar principles to the eye. The camera of a mobile phone, for instance, has a lens near the surface of the phone, and a silicon-based

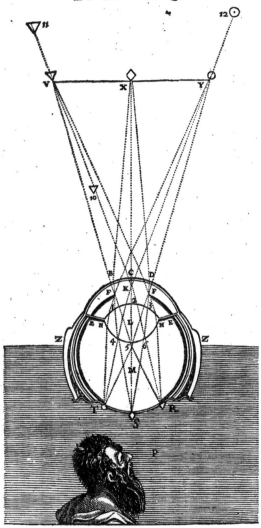

12. Descartes' experiment to show the image formed by an eye is upside down.

photodetector array inside the device. The mobile phone lens is often very small, and yet must provide images of sufficient quality that they are intelligible, and thus make sense when they are posted on Facebook. This requires that both the detector array and the imaging system are adequate to the task of producing a high-quality image. The quality of the image depends on two things—the size and the number of detectors in the array, and the ability of the optical system to create a sharp, undistorted image with all colours properly registered. That is, an image free from 'aberrations'.

The specification of the number of 'pixels' of the detector array is often used as a proxy for the quality of the image. A 24 megapixel camera (one in which the detector array contains 24 million sensors) is often considered better than an 8 megapixel one. A pixel can be thought of as the size of the image of a point object. If it is possible to resolve only a small number of points—because there are only a few elements in the detector—then it is hard to tell much about the object. The more pixels, then, the better. But only if the imaging system can produce a point image that is smaller than one detector element.

Limits to imaging

In the 19th century, the German scientist Ernst Abbe devised a simple rule for the minimum size of any image that was applicable to all imaging systems then known. Abbe's criterion says that the size (S) of the image of a point object is proportional to the wavelength (λ) of the light illuminating the object multiplied by the focal length (f) of the lens and divided by the lens diameter (D):

$$S = 1.22 \times \lambda f / D.$$

Thus, lenses with a big diameter and a short focal length will produce the tiniest images of point-like objects. It's easy to see that about the best you can do in any lens system you could actually make is an image size of approximately one wavelength. This is the fundamental limit to the pixel size for lenses used in most optical instruments, such as cameras and binoculars.

Designing and building imaging systems that can deliver high-quality images has been at the heart of many important applications of optics. Microscopes, for example, are used in applications from biological research to surgery. The earliest microscopes used very simple lenses—small, nearly spherical polished glass shapes that provided early experimenters like Robert Hooke in the 17th century with the means to explore new features of the natural world, too small to be seen by the unaided eye. His drawing of a flea, shown in Figure 13, was a revelation of the power of technology to enable new discovery.

Modern research microscopes are much more sophisticated devices. They consist of a multi-element compound lens that can form images with pixels very close to the wavelength of the illumination light—just at the Abbe limit. Figure 13 also shows an example of what is possible using a modern imaging microscope. It is a composite image of the nervous system of a fruit-fly larva, about to hatch, made by viewing light-emitting proteins located in the cells.

Abbe's criterion applies to all optical systems in which the image brightness is proportional to the brightness of the object. These are called linear systems. But it's possible to go beyond this limit by means of nonlinear systems, where the image brightness is proportional to the square or even more complicated function of the object brightness. A fuller explanation of these effects requires knowing a bit more about the wave model of light, which is the subject of Chapter 3.

Optical imaging systems of similar properties and complexity are used in another imaging application—the making of computer chips. Individual electronic circuit elements are extremely tiny.

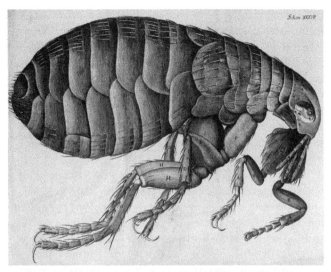

13. Hooke's diagram of a flea observed by means of an early microscope (top), and the nervous system of a fruit-fly larva (bottom) taken with a modern fluorescence microscope.

A wire connecting two transistors on a chip may have a diameter of only 250 nanometres (nm). (A nanometre is one billionth of a metre, or 10^{-9} m. For comparison, a human hair is approximately 100,000 nm in diameter.) The complex array of devices and connections is laid out on a silicon wafer by means of a process called lithography.

Essentially, the chip layout is drawn at a large enough scale to be visible to the human designers, then a demagnified image is projected on to the chip. The image is etched into a surface coating on the wafer, and a series of chemical processes then maps the image into real devices. The imaging system must be able to provide extraordinary resolution in the image—with pixel sizes of the order of the line size of the device. Maintaining this resolution over an entire wafer is a real challenge, requiring many lens elements properly designed to reduce all aberrations to an absolute minimum. An example of such a lens is shown in cross section in Figure 14, showing the multiplicity of lens elements and ray paths.

At the other extreme, both ground- and space-based telescopes for astronomy are very large instruments with relatively simple optical imaging components, often consisting of just one curved reflecting surface and a simple 'eyepiece' to adjust the rays so as to make best use of the available detectors. The distinctive feature of these imaging systems is their size. The most distant stars are very, very faint. Hardly any of their light makes it to the Earth. It is therefore very important to collect as much of it as possible. This requires a very big lens or mirror—several tens of metres or more in diameter. It is not practical to build lenses of this size, hence the ubiquity of mirrors in large telescopes. It is also necessary to look at distant stars for a long time in order to gather enough light to form an image. And this leads to another problem for ground-based telescopes: the atmosphere is not static. It changes in density with wind, temperature, and moisture. These fluctuations tend to make rays deviate from their course from star to telescope, causing the star to 'twinkle' as its light is deflected randomly onto and off the detector due to atmospheric turbulence.

One way to deal with this problem is to put the telescope outside the atmosphere, in space. The Hubble Space Telescope is an example. It produced spectacular images of distant stars, galaxies, and nebulae, showing extraordinary structures and movement in the far reaches of space. However, optical engineers have in the past two decades devised a clever way to deal with this problem

Light rays

14. Section of a lens used for photolithography of computer chips.
It consists of more than twenty different lens elements, and produces
images of 500 nm in size using light of less than half that wavelength.

for ground-based telescopes operating with visible light. What they do is to make the telescope mirrors in segments, the tilt of each segment being adjustable. Thus it is possible to 'steer' the rays hitting different parts of the telescope mirror so that they all hit the detector. If you can measure the deviation that a ray experiences as it traverses the atmosphere, then you can configure the mirror to compensate for that deviation. And that's what the engineers do. They measure how light from a guide star—an artificial light source in the upper atmosphere—is distorted and use that information to adapt the tilt of the mirror's segments. In this way images right at the Abbe limit can be produced. Space telescopes are still needed though, to probe the wavelength ranges, such as X-rays and UV, that are absorbed by the atmosphere, and several missions for new ones are planned by the North American Space Agency (NASA) and the European Space Agency (ESA).

Metamaterials and super lenses

For many years, optical scientists have been fascinated with the question of what makes a good optical system. Is there a lens that can form the perfect image of an object? This question has intrigued many great physicists, from James Clerk Maxwell in England in the 19th century to Victor Veselago in the USSR in the 20th century. Veselago thought about materials that behave in strange ways—in which light bends in a way opposite to that predicted by Snell's law. Snell's law is based on the positive refractive indices that are found in common 'normal' materials. Veselago considered materials with a 'negative' refractive index. Such materials can be made up of tiny structures that are each less than a wavelength of light in size. This kind of special construction gives 'metamaterials' such unusual optical properties. In particular, refraction takes place at the interface between normal and metamaterials such that the light rays bend in the opposite direction with respect to the interface than between two normal materials.

The strange refractive indices of these materials can be engineered to bend light rays in all directions, allowing the incoming rays which would normally be scattered by the object to instead be guided around it. Indeed, British physicist Sir John Pendry showed that it is possible to build an invisibility cloak using these designer materials.

Another of the unusual properties metamaterials possess is the ability to make perfect images of objects that are very close to a slab of the metamaterial. Even a flat surface is sufficient to make a lens, which makes them suited to viewing very tiny objects—so-called nanostructures, because they have sizes of the order of tens of nanometres. This is the 21st-century version of Hooke's technology, and will perhaps unleash a similarly fruitful era of discovery.

All of the imaging systems described in this chapter make two-dimensional renderings of objects. That's normally the way we experience and think about images—as flat pictures. But what if it were possible to conceive of a system that could make three-dimensional images? Remarkably it is, but that requires a deeper view of light itself, which we will consider in Chapter 3.

Chapter 3
Waves

Chapter 2 explored the idea of light as consisting of fundamental particles, moving along well-defined trajectories. This 'billiard ball' model, in which a light beam is a collection of individual, compact, and well-localized bundles of energy, stands in contrast to the alternative view, which is that light is a wave. This conception of light has been pursued in parallel with the particle view of light, although it took many years of discussion and experimentation for the wave picture to become fully accepted.

Unexplained phenomena?

If you look, when the Sun is shining, at a pool of water on which is floating a thin layer of oil, you will see coloured margins that map out the contours of the oil layer. It was this observation that stimulated historical thinking about light as a wave motion. Newton was one of the first to describe this effect, which provided a challenge to his particle model of light. Newton needed to make severe contortions to his model to account for this observation. Across the Channel, Newton's rival in optics and President of the Académie Française, Dutchman Christiaan Huygens, was able to explain this effect using a wave model of light. This turned out to be an altogether more elegant solution. So the ideas of wave and particle have been around simultaneously since the early days of the scientific study of light.

Other observations, such as those made by Francesco Grimaldi in the mid-17th century, gradually accrued evidence that did not fit with the particle model. Grimaldi saw that light rays deviated from straight lines when they passed through small apertures, such as a tiny hole in a screen. He noted that the light was diffused and that the edges of the beam were fringed with colour, especially pronounced for small objects such as a hair or a piece of gauze. He concluded that the striations seen when light was incident on a small or narrow object were evidence that the light had been bent from its original path as it passed the edges of these objects. If light really consisted of particles moving along straight rays, such solid objects would surely just cast a shadow, and not cause light particles to deviate into strange patterns.

Further, the problem well known to Newton and his contemporaries of the bizarre way in which light was refracted through certain materials, notably crystals such as calcite—a naturally occurring mineral—confounded explanation in terms of simple particles. An example of this behaviour is shown in Figure 15. The word LIGHT, written on a sheet of paper, is illuminated by a light bulb. Two pieces of calcite have been placed over each half of the writing. In the left half of Figure 15(a), two images of the word appear, displaced with respect to

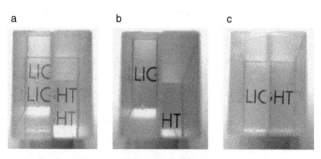

a   b   c

15. An illustration of birefringence. The image of the word LIGHT viewed through a pair of calcite crystals, using a. unpolarized light, b. vertically polarized light, and c. horizontally polarized light.

each other. In the right half, the two images are displaced in the opposite direction. The lower image in the left half and the upper in the right are just as one would expect from ordinary refraction of light reflected from the paper, seen through the crystal. But the second set of images appears to arise as if from a different refractive index. By placing a polarizer over the crystal, as shown in Figure 15(b) and 15(c), it is possible to isolate images formed by two different orientations of polarized light. Each polarization experiences a different refractive index. This is the phenomenon of *birefringence*.

These observations all pointed to features of light that demanded explanation. They are known respectively as *interference*, *diffraction*, and *polarization*. We will explore these phenomena in this chapter, developing a picture of light as wave motion.

Wavelength and wave frequency

What are the characteristics of wave motion? Waves are a form of undulation associated with a medium, for example water waves on the surface of a pond. These waves consist of the up-and-down motion of water molecules at the interface between the liquid and the air. The highest and lowest points of this motion become the peaks and troughs of the water wave, while the wave itself moves across the surface—that is, at right angles to the motion of the water molecules. For this reason it is called a *transverse* wave. Its speed depends on the depth of the water, among other things.

The circular surface waves that radiate away from the point at which a stone is cast into the water, as shown in Figure 16a, are a familiar effect. The distance between successive peaks is called the *wavelength* (Figure 16b), and the rate at which peaks hit the shore is called the *frequency* (Figure 16c). The product of these two quantities is the *speed* of the wave.

The puzzle for many centuries was what kind of undulations constituted light. It was assumed that some medium would need

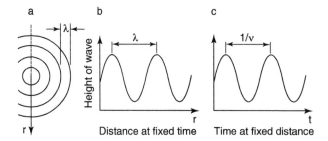

16. A circular wave on a water surface: a. contours of equal height, called wavefronts, b. wave height vs distance from centre at a particular instant, c. wave height vs time at a particular location.

to exist in order to support a wave. And since the speed of light is so large, this would need to be a very stiff medium. But a stiff medium would surely make it difficult for other bodies to move through it. For instance, because we can see distant stars, there should surely be some medium to support the propagation of light between the Earth and the star. Since the Earth moves around the Sun, the planet would be continually swept by a 'wind' as it travelled through such a medium. This enigmatic medium was called the 'aether', and it was not until the end of the 19th century that it was finally discarded as a useful concept.

So, what sort of wave is light? This was also finally answered in the 19th century by James Clerk Maxwell, who showed that it is an oscillation of a new kind of entity: the electromagnetic field. This field is effectively a force that acts on electric charges and magnetic materials. For example, a cloth charged with static electricity will attract dust particles to it. A magnet will be attracted to the door of a refrigerator. In the latter case you can feel this force as you place the magnet close to the door: the magnet accelerates towards the door unless an opposing force is supplied.

In both cases, a force pulls one object to the other, so that at some distance away from the cloth the dust particle feels this force due to the electric field generated by the charges on the cloth. And similarly for the refrigerator door, due to the magnetic field generated by the magnet. In the early 19th century, Michael Faraday had shown the close connections between electric and magnetic fields. Maxwell brought them together, as the electromagnetic force field. It turns out that in the wave model, light can be considered as very high frequency oscillations of the electromagnetic field. One consequence of this idea is that moving electric charges can generate light waves. I'll discuss this, and other methods for making light, in Chapter 5.

Interference

If two stones are dropped in the water next to one another, the resulting two sets of expanding circular waves collide. Some of the peaks are therefore higher, where the waves reinforce each other. But there are also lines of flat water, where there is no up-and-down movement of the water surface, even though the waves generated by both stones' disturbance of the water are passing through all points along these lines. The location of such lines is shown in Figure 17, along with the wavefronts—the locus of points on the peaks of the waves emanating from each source.

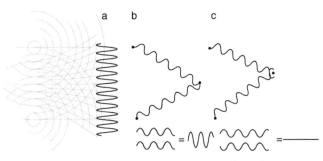

17. a. Two interfering circular waves on a water surface. The dashed lines are contours of equal phase. b. Constructive interference between paths of equal length. c. Destructive interference between paths differing by half a wavelength.

This phenomenon is known as interference, and it arises from the addition of the amplitude of two waves when they meet. If the peaks of the two waves coincide, the resulting peak is twice the size, or amplitude, of either wave. The waves are said to be in phase, and the interference is constructive. This is shown in the middle diagram of Figure 17. If, however, the waves are exactly out of phase, so that the peaks of one coincide with the troughs of the other, then the resultant wave has an amplitude of zero: the waves have 'cancelled' one another or interfered destructively. This is shown in the bottom diagram of Figure 17. It is immediately clear that such a phenomenon could not happen with particles, for how could two particles cancel one another?

It was the observation of interference effects in a famous experiment by Thomas Young in 1803 that really put the wave picture of light as the leading candidate as an explanation of the nature of light. Young's experiment was very simple, and very elegant. He took a candle as a light source, and placed it behind a screen in which there were two holes a very small distance apart. The light shining through these holes could be seen on a second screen some distance away. Now, if only a single hole was used—say by covering up the second—then the light made a single spot on the screen. When both holes were open, however, something marvellous happened—instead of just a spot of twice the brightness, the spot now had striations. These are generally nearly straight lines of zero intensity perpendicular to the line joining the centre of the two spots, as shown in cross-section in Figure 18 arising from the interference of the waves passing through the two holes. They are called 'Young's fringes', and are one of the key pieces of evidence for light as a wave motion.

How does interference explain the coloured 'fringes' observed by Newton in reflections from two surfaces very close together? What is needed for interference to occur is two waves, the relative phase of which (that is, the relative positions of the peaks of the two

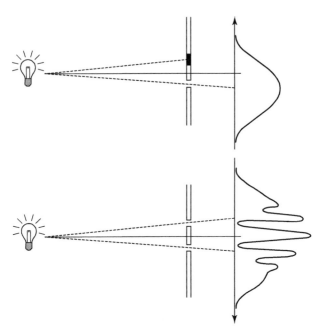

18. Thomas Young's experiment. The light from one slit produces a smooth intensity pattern. When both slits are open 'fringes' appear, characteristic of wave-like behaviour.

waves) can be adjusted. In Newton's experiment, interference occurs because a single incident beam of light is partially reflected at the two surfaces and thus split into two waves. If the distance between the two surfaces is equal to a single wavelength of the light, then the peaks of the two reflected waves coincide, and there is a bright fringe. If, however, the distance is just half of the wavelength, then the peak of one lines up with the trough of the second and there is a 'dark fringe' owing to destructive interference. Thus, when you look at the surface the bright and dark fringes map out the contours of the separation of the surface with a resolution of less than one wavelength. For green light, with a wavelength of approximately one-half of a millionth of a metre, or 500 nm, the separation can be determined to better than 250 nm—a precision of about 1/40th of the diameter of a human hair.

Of course, for a different wavelength the bright and dark fringes will occur in a different place, so that the surface shows fringes of colour when illuminated with white light. It is interference of light waves that causes the colours in a thin film of oil floating on water.

Interference transforms very small distances, on the order of the wavelength of light, into very big changes in light intensity—from no light to four times as bright as the individual constituent waves. Such changes in intensity are easy to detect or see, and thus interference is a very good way to measure small changes in displacement on the scale of the wavelength of light. Many optical sensors are based on interference effects.

Holography

Interference is also the means by which one can make true 3D images, that is, images that can be viewed from different angles and reveal different aspects of the object. These are different to the synthetic images in so-called 3D films, and are called holograms. Holograms are made by recording the full waveform of light scattered from an object. The sort of 2D images we're used to from photography encode only the amplitude of the waves. The phase information is lost. This is because detectors only respond to amplitude, so there is no way in the sorts of images we have looked at previously to extract phase. Nonetheless, it is the phase of the wavefronts scattered from an object that encodes its shape.

What interference enables is the encoding of phase into intensity, so that photodetectors can register patterns in which the full amplitude and phase information of the object wave is recorded. The principle is shown in Figure 19. The wave scattered from the object interferes with a reference wave that has a well-known

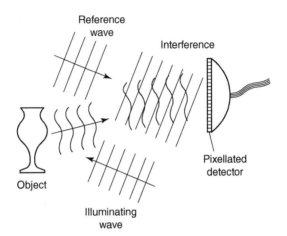

Reference wave

Interference

Pixellated detector

Object

Illuminating wave

19. A hologram is constructed by recording the interference fringes between a reference beam of light and one scattered from the object.

shape, often generated by a laser. The interference pattern is recorded on a detector or in light-sensitive materials: this is the hologram itself, invented by Denis Gabor in the middle part of the 20th century.

Viewing the image is a little more complicated than with an ordinary photograph. First, a reference beam illuminates the hologram and some of its light is scattered from the patterns encoded in the material. These scattered beams have the remarkable property that they reproduce the beams scattered from the original object, so that as your eyes receive them, it appears that the object is reconstructed in front of you. Moving around the image reveals different aspects of the object, because the beams scattered from those different parts encode different information.

Holograms can also be computer generated and embossed in metals or other materials. The idea is that the surface shape mimics the interference pattern of the reference and object waves, so that raised parts correspond to bright fringes and low parts to dark fringes. Again, illuminating this with a reference wave causes scattered light to imitate the wavefronts of the chosen object. This kind of hologram is used as a security device, including on some banknotes (such as the UK £20 note, which has a strip of holographic images of the 18th century Scottish economist Adam Smith on it), because they are hard to make and require advanced engineering to copy.

Limits to imaging, redux

The wave picture of light also explains why we can't see infinitesimally small objects using a microscope, just as Abbe realized. Very tiny things—down to about half of one micron (one millionth of a metre, or about half a wavelength of visible light) can be seen using an ordinary light microscope. Much more sophisticated methods are required to see even smaller things. The reason is that the wave nature of light puts a lower limit on the size of a spot of light.

I noted previously that when two light beams meet, they interfere to give regions of zero amplitude—dark fringes. The distance between these fringes is actually determined by the angle at which the two beams intersect. If the angle is very large, the fringe spacing is small; if the angle is small, the fringe spacing is larger. The smallest possible separation of the fringes is one-half wavelength, or about a quarter micron for visible light.

Now, if this fringe pattern is recorded as a hologram, then when it is illuminated again with the reference beam two beams will emerge in the directions of the light beams that were used to record the interference pattern. In order to see such a fringe pattern with a microscope the lens must capture both of these beams to form an image of the fringes. If this didn't happen, say because the lens could only capture one beam, then there would be no interference fringes in the image.

This is the physical basis of Abbe's criterion that I introduced in Chapter 2: the maximum angle between two beams captured by the lens of an imaging system sets the minimum object size that can be resolved. It's straightforward to see then that the best any lens system can do is to see object sizes of about the wavelength of the illumination. Thus conventional light microscopes can see very tiny objects, about fifty times smaller than a human hair, but not much smaller than this. They can be used to look at biological cells, but not the cell nuclei, for instance.

Super-resolution imaging

Optical scientists and engineers have found a number of very clever ways to get around the object size limits posed by conventional microscopes, so that they can see inside cells, or view objects that are more than one hundred times smaller than the wavelength of light. These instruments make use of new materials and processes, such as the ability to attach nanometre-scale particles to the objects you wish to see, or to insert molecules that emit light into cells. These emit light (they fluoresce) at a long wavelength when they are illuminated by a beam of light at shorter wavelengths. Since they are much smaller than the resolution of the microscope lens, the resulting image is a spot the size of which is entirely limited by the microscope optics, according to Abbe's formula. But the exact centre of this image can be located very precisely by using a camera to take a long look at the fluorescence of the attached nanoparticles and determine the position of the maximum intensity of the spot. This approach is called photo-activated localization microscopy, or PALM, and was invented by Eric Betzig in the US. It has revolutionized live cell imaging, allowing faster acquisition and precise depth resolution over a wide field of view.

Another method for measuring small structures in larger fluorescent objects is to shine a second annular beam on them that causes the objects illuminated by this beam to have their fluorescence extinguished, so that the remaining fluorescence can be located precisely by the same approach as described previously. This method is called stimulated emission depletion microscopy, or STED, and was invented by Stefan Hell in Germany. I will describe the process of stimulated emission in more detail in Chapter 5. These novel ways of making high-resolution images that enable intracellular structures to be imaged have had enormous impact in biology and medicine. Indeed the significance of this impact was recognized when the 2014 Nobel Prize in Chemistry was awarded to Betzig and Hell (together with W. Moerner).

The process identified by Abbe works in reverse, too. When applied to the illumination of a sample by means of a microscope lens, it says that a light beam cannot be focused to a spot much smaller in diameter than a wavelength. Again, the tightness of the focus depends on the range of angles that the lens can produce on the side that faces towards the object: the broader the range of ray directions, the tighter the focus of the light.

The relationship between the range of angles between interfering beams and the size of the fringe structures turns out to be a very fundamental property of waves. This idea was quantified by Joseph Fourier, a French scientist of the early 19th century, who provided a detailed mathematical analysis of light wave propagation. Fourier's theorem says in simple terms that the smaller you focus light, the broader the range of wave directions you need to achieve this spot.

Diffraction

This explains another feature of light beams—they gradually diverge as they propagate. This is because a beam of light, which by definition has a limited spatial extent, must be made up of waves that propagate in more than one direction. The idea can be tested using a laser pointer. The beam emitted by the laser itself is about 1 mm (one thousandth of a metre) or so in diameter. When it reaches the screen, it is about 60 mm (sixty thousandths of a metre) in diameter. And if it were sent further away, say to the Moon (approximately 400,000 km), then it would be about 240 km in diameter! This phenomenon is called diffraction.

Diffraction has some interesting applications in determining the shape and symmetry of structures. For instance, when you shine a beam of light on a screen with small holes in it, of diameter comparable to the wavelength of the light, the light diffracts through the apertures, spreading in inverse proportion to the size of the aperture. These diffracted beams interfere some distance away from the screen, and the resulting interference fringes—the so-called diffraction pattern—tell something about the size and relative location of the apertures. For instance, if they are in a regular array, then the diffraction pattern will also show regularities. The advantage of using such patterns to measure this type of object is that you don't need to have very expensive or complicated lens systems or detectors close to the object—you simply look at the pattern when it has naturally expanded due to diffraction.

Now, imagine that the screen is replaced by a transparent solid material, say a crystallized protein structure. The 'holes' are replaced by the atoms in the protein molecule, which are very small indeed and are connected to one another by bonds in the molecule that are about one tenth of a billionth of a metre (0.1 nm) in length. If light with a wavelength of about this size illuminates such a structure, then the light will be diffracted. The actual structure of the molecule itself can be determined from the diffraction pattern. This is the basis of *X-ray diffraction*. As noted in Chapter 1, it was famously used in the effort to find the structure of DNA and is now a very common tool in biochemistry, used regularly for finding out the structure of new molecules that might be useful in developing drugs for example. It requires a bright X-ray light source, as well as a means to make crystals out of the molecules. Figure 20 shows a diffraction pattern from a crystal of bovine enterovirus.

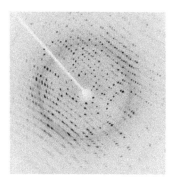

20. An X-ray diffraction pattern of a protein crystal taken using a modern synchrotron X-ray source.

It's clear that if you want to transmit light over long distances, then diffraction could be a problem. It will cause the energy in the light beam to spread out, so that you would need a bigger and bigger optical system and detector to capture all of it. This is important for telecommunications, since nearly all of the information transmitted over long-distance communications links is encoded on to light beams.

Guided waves

The means to manage diffraction so that long-distance communication is possible is to use wave guides, such as optical fibres. A wave guide is a structure that has a carefully designed profile for the refractive index. For example, the index variation across an optical fibre is arranged so that the 'core', with a diameter of a few millionths of a metre, has a higher index than the surrounding 'cladding'. The light is confined to this higher index core and moves along the fibre without diffracting, remaining the same size over very long distances—across the Atlantic Ocean via undersea cables, for example. This means of controlling light is the basis of a wide-ranging optical information infrastructure, from communications to sensors.

Polarization

The final important feature of the wave model is the characteristic of polarization. Recall that in transverse waves the undulations occur in a direction at right angles to the direction of propagation of the wave. Importantly there are two such directions for these undulations.

Consider waves on a string, for instance. If you move one end of a rope up and down rapidly, you can see the undulations move along the rope. A similar thing happens if you move the end left and right equally rapidly. Both vertical and horizontal oscillations are at right angles to the motion of the wave down the rope. The waves are said to be 'transverse'.

Light waves behave similarly. For example, horizontally polarized light has an electric field that oscillates in a horizontal plane (relative to, say, an optical bench). Similarly, the field of a vertically polarized light beam oscillates in a vertical plane. (There are more complicated forms of polarization too, but these will suffice for now.) The phenomenon of birefringence can be explained by noting that a crystal is a structure of atoms in very specific and highly repeatable units. These 'unit cells', consisting of a few atoms, may themselves be asymmetric. Depending on whether the light is polarized along the long axis or the short axis of the unit cell, it will experience a different refractive index, and thus will be deviated by different amounts as it propagates through a block of such material.

One of the well-known ways in which polarization of light is exploited is in sunglasses. Some of these use lenses made of plastic material (for example, a sheet of Polaroid plastic) that acts as a polarizer—that is, an element which transmits light of a particular polarization (say vertical) and absorbs that of the orthogonal polarization (horizontal). Polaroid is made from rugby-ball shaped molecules that are aligned and 'frozen' in the plastic polymer. These molecules preferentially absorb light that is polarized along the axis of alignment of the molecules. Since generally sunlight scattered from objects has a random polarization (roughly 50 per cent of each polarization direction), then filtering out light of one polarization effectively reduces the brightness of the scene by half. Further, polarized sunglasses reduce glare: that is, light reflected from shiny flat surfaces, such as the hood and windscreen of cars. These surfaces tend to reflect a higher proportion of light polarized in a direction parallel to the surface (an effect discovered in the 19th century by Sir David Brewster and now named for him). Such reflections are blocked by spectacles that are constructed in the manner noted here, making for clearer vision of the road.

Transparent birefringent materials can also change the polarization of light without absorbing the light. This is because the speed of light depends on the direction of polarization with respect to the 'orientation' of the material. Some materials, such as ordinary glass, have no special orientation: you can rotate the material with no change in the effect on a light beam. As noted above, atoms in birefringent materials are arranged in such a way that there is a preferential direction—the symmetry axis—along which atoms respond differently to light. That is, light polarized along the symmetry axis will go slower (say) than light polarized perpendicular to the axis. Now, imagine that the light is polarized at 45 degrees to the symmetry axis. We can consider that half of this light is then polarized along the symmetry direction and half polarized at right angles to this direction. If the latter is slowed down by a sufficient amount, then the light emerging at the other face of the material can be polarized at −45 degrees. Thus the direction of polarization has been 'rotated' by 90 degrees.

Some birefringent materials can be used to control the state of polarization, by actively adjusting the orientation of the alignment axis of the molecules, say, using an applied voltage across the material itself. An example of this is a class of materials called liquid crystals (LCs), which consist of elongated molecules. The orientation of the molecules in an LC can be controlled by applying a voltage across them. Other materials become birefringent when forces or stresses are applied, because the force causes the molecules to rotate or the atoms to change alignment. This phenomenon enables the construction of force sensors by monitoring the state of polarization of light at the output of the optical sensor.

Placing a piece of birefringent LC between two polarizers also allows control of the light intensity by means of an electrical connection. Applying a voltage reorients the molecules, thus changing the refractive index seen by a polarized light beam. If a polarizer is placed after the LC then depending on the applied voltage a greater or lesser amount of light will be transmitted through the final polarizer. An array of such 'cells', each driven by a separate electrical signal, can be used to form a display where each cell is a single pixel. This is the basis of a liquid crystal display (LCD), and is often used for computer display screens and television sets.

In fact, such displays can be used to show 3D movies. The illusion of depth in these movies derives from the stereography of human vision. Each of our eyes sees a scene from a slightly different location, since they are a few centimetres apart in our skulls. The two images are combined in our brains to give us a perception of depth.

This illusion is reproduced using 3D glasses by means of polarization. Two images are projected on to the display or screen. Each of these is generated using light of a specific polarization, and each is shot from a slightly different vantage point. The 3D glasses are polarizers set at different orientations that allow transmission of one scene to the left eye and one to the right, each completely blocking the alternate image that has the 'wrong' polarization. Thus we have a sense that the scene is as we would perceive it in the natural world—that is, the illusion of a three-dimensional arrangement of objects and people.

The successes of the wave model of light have been stunning, allowing us to understand some important characteristics of light and to use this understanding to build new technologies. And the successes of the ray picture of light have been equally amazing. Yet it is certainly puzzling that two apparently very different views of what light is should be necessary. It is to this conundrum that I now turn.

Chapter 4
Duality

The two different views of light, as a particle and as a wave, both contain insight and value. They have each enabled both new understanding of the natural world and the development and design of new technologies. Yet they appear to be vastly different in their conception of what light actually is. On the one hand, the particle model views light as a localized entity, a bundle of energy, that moves along a well-defined trajectory. On the other hand, the wave model describes light as a diffuse entity, permeating through space with no connection to the motion of solid things. How can these two pictures possibly refer to the same thing? This dilemma was recognized early on by Huygens and his contemporaries, but the two views remained in tension, as alternative descriptions of light, until the 19th century.

When Maxwell developed his theory of electromagnetic fields, he was able to use this to explain the properties of light as wave motion of those fields, as we saw in Chapter 3. This triumph of reasoning appeared to confirm the experiments of Thomas Young and Auguste Fresnel (described in Chapter 3) by providing an explanation of two fundamental phenomena, interference and diffraction, that did not easily fit within the particle model. Yet the concept of trajectories remained, and still remains, an extraordinarily powerful one for the analysis and design of optical systems. So there's an uneasy truce of these two pictures—a dualism within classical physics—that requires some consideration. How can they be reconciled?

Looking at trajectories again

In the 17th century the Frenchman Pierre de Fermat proposed an ingenious formulation of refraction that was very different from that of Snell. Recall that Snell's law deals with the change of direction of a ray of light at an interface between two transparent media. The ray, defined by the direction in which it is travelling towards the interface and the point at which it hits the interface, has its direction altered by an amount proportional to the ratio of the refractive indices of the two materials. It is only the local properties of the ray and interface that are important. Snell's law applies at each point along the trajectory, as if the ray is 'feeling' its way along, adjusting direction when it encounters a new interface.

Fermat's conception was radically different. He argued that one should define the trajectory in terms of starting and ending points, as shown in Figure 21. He suggested that the question to

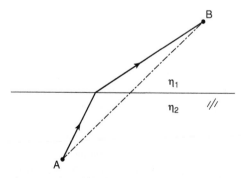

21. **Fermat's conception of a light ray as a path of least time connecting the start and end of the trajectory. The ray crosses an interface between two optical media in which light moves at different speeds.**

ask is: what is the path that the light takes to traverse the space between the two points? He proposed that it should take the path that minimizes the time of flight between the two points. That this gives the same answer as Snell is remarkable and profound. Fermat's 'principle of least time' suggests that the light considers the overall picture of the situation, and that the notion of a ray is one that takes into account both the initial and final positions and directions as well as everything in between. The contrast with the local model of a particle reacting to its immediate environment is telling.

This idea was taken up by the German natural philosopher Gottfried Wilhelm von Leibniz, who was Newton's contemporary and antagonist. Leibniz was impressed by the holistic picture of the process described by Fermat, and the concept of 'optimization' that it implied: a ray explores the whole of space and picks just that path that will minimize its transit time between the specified beginning and end points. He developed the mathematical tools for analysing this idea—the calculus of variations—by which the effects small changes on a trajectory would have on the time taken to traverse the modified trajectory could be calculated. Leibniz recognized the importance of the notion that Fermat's principle provided: the movement of light from one point to another defines an 'optimal' trajectory.

Indeed, so taken was Leibniz by this concept of optimization that he elevated it to a teleological principle: that the world itself, in all its aspects, was on the optimal trajectory between a starting point and a finishing point. The contradictions inherent in such a position, when applied outside of the realm of science, were ably lampooned by Voltaire in his novel *Candide*, where Leibniz's ideas are put into the mouth of Dr Pangloss, who insists disasters both natural and man-made were nonetheless evidence that this is the 'best of all possible worlds'.

Connecting waves and rays

Nonetheless, Leibniz's mathematical ideas proved to be very fruitful. They were taken up by the renowned Irish mathematician William Rowan Hamilton in the 19th century. He showed formally how the idea of a wave can be allied to that of a collection of particles. Waves can be defined by their wavelength, amplitude, and phase (see Figure 15). Particles are defined by their position and direction of travel (see Figure 5), and a collection of particles by their density (i.e. the number of them at a given position) and range of directions. The media in which the light moves are characterized by their refractive indices. This can vary across space. For example, at the interface shown in Figure 20 there is a step change in the refractive index across the boundary between the two media.

Hamilton showed that what was important was how rapidly the refractive index changed in space compared with the length of an optical wave. That is, if the changes in index took place on a scale of close to a wavelength, then the wave character of light was evident. If it varied more smoothly and very slowly in space then the particle picture provided an adequate description.

He showed how the simpler ray picture emerges from the more complex wave picture in certain commonly encountered situations. The appearance of wave-like phenomena, such as diffraction and interference, occurs when the size scales of the wavelength of light and the structures in which it propagates are similar. Thus you see diffraction patterns arising when the object that the light hits is a few microns in diameter, or has a very sharp edge, such as the delicate structure in a bird's feather, or a butterfly wing. Otherwise, as in the case of a camera lens, the trajectory provides a sufficient description, since the refractive index is uniform throughout the glass of the lens itself.

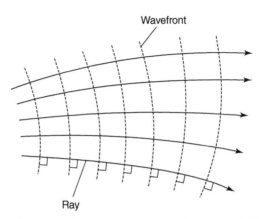

Wavefront

Ray

22. Hamilton's idea of rays as connecting wavefronts—thus joining the two primary conceptions of light.

Further, Hamilton showed that Fermat's trajectories related directly to a property of the wave—the wavefronts. These are the loci of points at which the wave has the same phase at each point in space. For instance, when you see the ripples on the surface of a pond after a stone has been cast into it, the circular patterns are just these wavefronts. They are the places on the surface where the wave 'peaks' (or troughs) at a given instant of time. Now, what Hamilton noted was that rays could be considered as lines that intersected the wavefronts at right angles, as shown in Figure 22, thus connecting adjacent wavefronts by a well-defined trajectory.

Hamilton's 'optical analogy'

This remarkable result suggested another profound comparison— Hamilton's so-called 'optical analogy'. What he noted was that the well-known formulation of mechanics—the motion and position of solid bodies of matter—was based on the idea of trajectories. The idea that these may also be in some sense 'optimal' had been considered by Pierre Louis Maupertuis in the 18th century.

Maupertuis had formulated a way to evaluate the optimal value of a quantity called the 'action'—essentially the velocity of the body multiplied by the distance it moves (and times its mass)—along the body's trajectory.

He argued that the action should be minimal for an actual trajectory between two points, just as in Fermat's argument that the time taken to traverse a light ray should be minimal. Maupertuis' 'principle of least action' is very similar in concept to Fermat's 'principle of least time'. Indeed, Leonhard Euler, a Swiss mathematician, showed how to use Leibniz's calculus to derive Newton's famous equations of motion from Maupertuis' principle. Thus Euler connected a description of a trajectory in terms of a particle sensing its way through its environment to one in which the whole of space between the specified starting and finishing points influences the path.

What Hamilton did was to find equations that encapsulated the variations in action in terms of a simple description of the specific environment in which the body was moving. And this equation turns out to have a very similar form to the one he found for describing the trajectories of light rays (for which the environmental description is just how the refractive index changes with position in the medium). So there is a hint of a latent analogy between the trajectories of solid objects and a fictive wavefront: perhaps all bodies might have both particle-like trajectories and wave-like properties? Indeed, Hamilton's equation, and his eponymous function, turns out to be very important in thinking about the next big step in understanding light—quantum mechanics.

Unsolved puzzles

This was by no means the only hint of a new opportunity for science. About this time, towards the end of the 19th century, light still offered a few puzzles that were unexplainable in terms of the prevailing models of its properties, even with the reconciliation that Hamilton had provided. Two of the most important of these were: the colour of hot objects (including the Sun), and the colour of different atoms in a flame.

When things are heated up, they change colour. Take a lump of metal. As it gets hotter and hotter it first glows red, then orange, and then white. Why does this happen? This question stumped many of the great scientists of the time, including Maxwell himself. The problem was that Maxwell's theory of light, when applied to this problem, indicated that the colour should get bluer and bluer as the temperature increased, without a limit, eventually moving out of the range of human vision into the ultraviolet—beyond blue—region of the spectrum. But this does not happen in practice.

The second example arose from the study of light emitted by atoms, pioneered by a Swiss schoolmaster, Johannes Balmer. We'll look at this mechanism in more detail in Chapter 5, but the important feature of the light is in the distribution of colours in its spectrum, shown in Figure 23a. In this respect it is very different from sunlight (the Sun is a good example of a hot body), which has the familiar 'rainbow' spectrum, shown in Figure 23b, consisting of all colours continuously from red to violet (and beyond at each end—just not visible to us). By contrast, a collection of atoms emits a discrete set of colours—a set of spectral 'lines' of particular wavelengths—specifically associated with the internal structure of the particular atom involved.

Both of these phenomena required a radical revision of thinking about light, because they could not be explained within the contemporary models of wave motion and atomic structure.

Max Planck, working at Humboldt University in Berlin in the late 19th century, first came up with an idea to explain the spectrum emitted by hot objects—so-called 'black bodies'. He conjectured that when light and matter interact, they do so only by exchanging

a

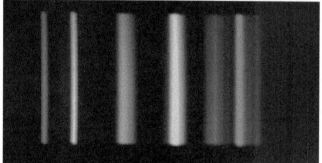

b

23. **Spectrum of light emitted from a. the Sun (a 'black body'), and b. a neon lamp. The former has continuous band of colours, whereas latter shows the discrete lines of particular colours that are a 'fingerprint' of the neon atoms.**

discrete 'packets', or quanta of energy. Planck recognized the revolutionary nature of his idea, and was therefore reluctant to infer too much about light itself from it, though it would eventually dramatically change our view of what light is. His notion revived the idea of light as a stream of particles—discrete objects that carried a fixed amount of energy that could be absorbed by atoms or emitted by them.

This seemed like a retrograde step: the wave model of light could explain all of the hitherto observed effects, and it was clear from Hamilton's work that even trajectory-like behaviour, previously the most obvious evidence of a particle-like entity, emerged from a wave model of light in certain common situations. So the idea of a particle of light appeared not to be necessary. Surely it was simply a calculational 'fix', thought up to get out of a tight spot, and would eventually be replaced by a more reasonably consistent picture of light. However, combined with Balmer's observations, this conjecture was set to radically change physics.

In the years immediately after Planck's suggestion, Albert Einstein used the idea of discrete exchanges of energy between light and matter to expose another piece of physics that had eluded explanation—the photoelectric effect. This effect is seen when light shines on a piece of metal. Some electric charges—electrons, in fact—are ejected from the metal. The speed with which the electrons are ejected depends on the wavelength of the light. The light must be sufficiently 'blue', that is have a short enough wavelength, in order to see any electrons emitted at all. As it gets bluer and bluer, the electrons are ejected with more and more energy, and thus higher and higher speed.

Einstein explained this by noting that the electrons required a minimum amount of energy to escape the clutches of the metal, and by assigning discrete amounts of energy to a particle of light—the photon—proportional to the frequency of the light (the constant of proportionality being known as Planck's constant, h), so that when the frequency of the light is high enough (and thus the wavelength short enough) the light, when it is absorbed, can provide enough energy for an electron to escape. His model suggested that the origin of the discrete exchange of energy between light and matter that is central to Planck's notion arises from the actual discrete character of light—a full revival of the particle model.

This idea chimes nicely with Balmer's observations of discrete line spectra of light emission from atoms. But a full explanation of this phenomenon clearly requires a bit more thought about why atoms would deliver light in such packets. The key idea came from Niels Bohr, a Danish physicist working in Manchester. He suggested that the reason light was emitted as discrete packets of energy was that the atoms themselves could only exist in certain configurations. He imagined atoms as analogous to tiny planetary systems: electrons in orbit around a central nucleus. The electron can 'jump' between two stable orbits, emitting or absorbing light as it does so (depending on whether the jump is to an orbit of lower or higher energy). The character of these orbits—or quantum states—depends on the details of the atom: how many electrons and the size of the nucleus. Therefore the energy given or taken up when an electron moves between two quantum states is a signature of the particular atom itself. So when the energy of a light particle—or photon—matches that of the difference in energy of two quantum states of the electron in an atom, then absorption or emission is possible. Bohr's ideas explained Balmer's observations neatly, and added some weight to the idea of light beams as a collection of discrete particles.

All these developments had the potential to undermine the wave model of light that had been so strongly affirmed by Maxwell's theory. They went beyond even Hamilton's reconciliation of trajectories and wave motion, since they appeared to be fundamental, not simply the result of an approximation about size and scale. Thus they reopened the question of the nature of light once again.

In 1908, George Taylor, working in Cambridge, performed Young's double-slit experiment with exceptionally feeble light—so weak that on average there was less than one photon in the apparatus at any time. Yet he still saw interference fringes. That outcome is very strange. If we think of one path from the light source to the detector as being via one slit, and a second path via the other slit, then there are two ways that the photon can get from source to detector. The evidence of interference fringes posed a dilemma that caught the attention of the leading scientists of the day. Bohr captured the difficulty, noting that we would 'be obliged to say, on the one hand, that the photon always chooses one of the two ways, and on the other that it behaves as if it passed through both'. Even single particles can exhibit wave-like behaviour.

Waving altogether

As you might imagine, it took a truly radical thought to figure a way out of this conundrum. That thought occurred to Paul Dirac, a physicist working at Cambridge in the 1920s. Dirac suggested that the fundamental property of light was that it was both a particle and a wave at the same time. Now, this might appear to you as simply sophistry, a logical trick that resolves nothing. But there is much more behind it than a slogan. What Dirac did was to develop a quantum mechanical version of Maxwell's theory of electromagnetic fields. Using this he was able to show that, if you measured these 'quantum fields' using a set-up like Young's double-slit interferometer, you would see interference effects characteristic of wave-like behaviour. Whereas, if you simply measured the intensity of the light, you would be effectively counting the number of photons in the beam.

This turned out to be a very profound step. It set the quantum field up as the fundamental entity on which the universe is built—neither particle nor wave, but both at once; complete wave–particle duality. It is a beautiful reconciliation of all the phenomena that light exhibits, and provides a framework in which to understand all optical effects, both those from the classical world of Newton, Maxwell, and Hamilton and those of the quantum world of Planck, Einstein, and Bohr. But the cost is a perplexing and non-intuitive entity at the heart of nature—a quantum field—of which light is but one example.

The radical idea that light was both wave and particle stimulated some major new ideas. For instance, Louis de Broglie suggested that if this kind of dualism existed for light, surely it should for all other things too. So, material bodies that we normally had considered only as collections of particles should also have 'wave-like' properties, taking the next step beyond what Hamilton had considered. He even defined what the wavelength should be (now called the de Broglie wavelength λ_{dB})—proportional to the inverse of the particle's momentum (i.e. its mass and velocity, the constant of proportionality again being Planck's constant):

$$\lambda_{dB} = h/mv.$$

This suggests that to look for such effects you should use either very light particles or very cold ones that are moving very, very slowly. Such effects can be observed. Figure 24 shows an interference pattern made using a double-slit-like interferometer, but using molecules instead of light. The implications of this are mind-boggling. If you think of a molecule as just a very light particle, then you cannot explain the pattern, because you consider that it must have passed through one slit or the other. However, the idea that a particle with mass could be so delocalized as to have passed effectively through *both* slits to interfere with itself is astounding and beggars belief.

If material objects also behave as waves, then, Erwin Schrödinger conjectured, surely there must be a wave equation describing their behaviour. Where to start looking for such a thing? He took the equation that Hamilton had developed, by means of his optical analogy, to describe how the 'action' of a particle evolved. A simple addition to this equation, involving Planck's constant, moulded it into one that described wave motion. This was the origin of Schrödinger's famous 'wave function'. The wave function has a number of properties very analogous to optical waves, including interference and diffraction, but nonetheless refers to things that in the language of pre-quantum physics are palpably particles

24. An interference pattern made using molecules passing one at a time through a tiny version of Young's apparatus—two very small slits separated by billionths of a metre.

possessing mass and weight. As a consequence, there are still questions as to what the wave function actually describes. Is it the actual particle itself, or is it a sort of shorthand for what we know about the particle?

An important quantity for light is its intensity, proportional to the square of the amplitude of the field. This is related directly to the density of photons in the light beam. Similarly, the square of the wave function is related to the density of particles at a particular point in space at a particular time. But it is impossible to say for certain that a particle occupies a specific position at a specific instant. This indeterminacy seems to be a very fundamental property of the world, and it is related deeply to the fact that quantum fields lie at the heart of things.

Nothing is something

Another consequence of this fact is that 'nothing' is actually something. That is, the complete absence of matter (e.g. electrons or atoms) or light (i.e. photons) still has measurable properties. This void is called the 'electromagnetic quantum vacuum', and is the state of the universe from which all extractable energy has been removed. Yet it is a seething mass of activity, consisting of fluctuating fields but containing no photons whatsoever. What is even more surprising is that the quantum vacuum has observable consequences. How can this 'nothing' give rise to an effect we can see?

We've seen that light can be thought of as a wave motion of the electromagnetic field. Imagine this field as ripples on the surface of the sea. These would buffet about any boat, but would not move the boat up and down as a well-defined wave might or push it along. On average, the boat does not move, but it nonetheless rocks back and forth. Now imagine the same thing for a charged particle like an electron. It 'feels' the random changes in the electromagnetic vacuum, being buffeted by these. If the electron is bound in an atom, this buffeting is revealed as a shift of the energy of the quantum states that the electron may occupy. Since the frequency of the photon that the atom may absorb depends on the difference in energy of two such states, these changes can be seen in a change of the colour of light that the atom may absorb. The changes are tiny—a shift of less than one billionth of the wavelength of the light—but measurement techniques for frequencies are so precise that such changes can indeed be determined. The first person to do this, Willis Lamb, working in New York in the 1950s, won a Nobel Prize for showing this frequency shift, which is now named for him.

The dual identity of light has numerous facets. Even in the pre-quantum world, the dichotomy of ray and wave demanded a resolution. That came about by understanding the nature of the wave motion that light embodied, and the scale and nature of objects with which it interacted. Particle-like behaviour—motion along a well-defined trajectory—is sufficient to describe the situation when all objects are much bigger than the wavelength of light, and have no sharp edges. Quantum mechanics puts a new twist on this duality. Light acts as a particle of more or less well-defined energy when it interacts with matter. Yet it retains its ability to exhibit wave-like phenomena at the same time. The resolution is a new concept: the quantum field. Light particles—photons—are excitations of this field, which propagates according to quantum versions of Maxwell's equations for light waves.

Quantum fields, of which light is perhaps the simplest example, are now regarded as being the fundamental entities of the universe, underpinning all types of material and non-material things. The only explanation is that the stuff of the world is neither particle nor wave but both. This is the nature of reality.

Chapter 5
Light matters

How do you generate light? There many different ways: common light bulbs, based on glowing metal filaments, as well as 'fluorescent' tubes; laser pointers; tell-tale lights on electrical equipment, from toasters to car dashboards; sunlight and starlight; even, for those in the northern and southern extremities of the planet, the *aurorae*; fireflies and glow-worms, and the phosphorescence in the wake of boats. What is the means by which these very different things all generate the same thing: light?

The key is that they all involve matter. More specifically, they involve electric charges moving about. When these charges accelerate—that is, when they change their speed or their direction of motion—then a simple law of physics is that they emit light. Understanding this was one of the great achievements of the theory of electromagnetism. An electric field has its origin in an electric charge, such as an electron in an atom. The electric field, attracting oppositely charged particles, such as protons, extends throughout all space, although it gets weaker quite quickly as you move away from the electron. As I noted in Chapter 3, this is the force that arises from static electricity.

Oscillating atoms and bending electrons

Now say the electron moves with a sudden jerk. The field surrounding it must move also, since it is linked inextricably to the electron. The change in the field is illustrated in Figure 25, as a 'kink'. A remote proton will notice that the electron has moved only when the change in the field that arises from the electron movement has had time to make its way to the proton. This will take time since the kink, and therefore the information that the electron has moved, propagates at about the speed of light. When the change does get to the proton, it will move according to whether the electron moved closer to it (thus making the field experienced by the proton stronger, so that it feels a stronger force) or further away (leading to a weaker field and a weaker force).

Now say the electron were to move back and forth. Then changes would occur in the field surrounding it in synchrony with this oscillation and propagate away to the proton, which would then be induced to oscillate in turn as it received this information. But an oscillating electric field (and associated magnetic field, but that need not bother us here) is exactly what we mean by light.

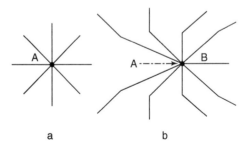

a   b

25. Lines in the electric field of a. a stationary electron and b. an accelerating electron. When the electron accelerates from A to B, the changes in the field—the 'kinks' in the field lines—propagate away from the electron at the speed of light.

Since a hydrogen atom—the simplest one—consists of a single electron and a single proton, we can understand from this picture how atoms can generate light. First, let's think about what happens when light irradiates an atom sitting quietly on its own. The light forces both charged particles—the electron and the proton—inside the atom to move. But since the electron is much lighter than the proton, it moves more easily with a given application of force, and we can consider its motion with respect to the more or less stationary proton. In fact, the electron oscillates at the frequency of the light's electric field, being alternately accelerated or decelerated as the electric field varies.

This process is a bit like pushing a child on a swing. The best way to get the swing to oscillate is to push on it in synchrony with the natural oscillation period—a gentle push each time the swing reaches the bottom of its motion. Even so, it takes effort to get the child up to a height that thrills him or her. The child experiences maximum acceleration at the apices of the swing's oscillation, and maximum speed as she passes under the point of suspension. So it is with the atomic electrons—the energy of the light beam is absorbed by the atom, and converted into the motion of electrons.

Let's say you now stop pushing on the swing. What happens? The child gradually swings in arcs of lesser and lesser amplitude to rest. Again, so it is with the atom. The electrons gradually stop oscillating, and give up their motional energy by reradiating light. This is the process of light emission, and is the basis of many light sources, such as neon signs, fluorescent lights, and laser pointers.

Now in this picture I have assumed that some light beam itself is used initially to get the atomic electrons oscillating. But in a sense that begs the question of how one generates light in the first place. In fact it is possible to use other means to 'excite' the atoms. First, one can simply heat up the material. This is what is done in an ordinary light bulb, where by passing an electric current through a metal filament the metal heats up to a very high temperature—several thousand degrees. As the material heats up, the electrons start to jostle and collide with the atoms and with one another more and more, and this both excites the atoms and causes the electrons to accelerate and decelerate rapidly. This produces a very broad range of colours that depends only on the temperature to which the material is heated, and not upon the particular type of atoms of which it is made.

Electricity may also be used in other ways. In light-emitting diodes (LEDs) for example, as used in displays, an electrical current—or flow of electrons through the material—can be captured directly by an atom, so the light is generated much more efficiently than in a thermal source. A fluorescent tube also uses an electrical current to directly excite atoms, only this time in a nebulous gas with which the tube is filled. Finally, many different chemical or biological reactions can release energy, some of which leaves the atoms or molecules in the form of light. This is the origin of the light given off by fireflies, for example.

We noted previously that acceleration has two parts to it: it can mean a change of speed, as we have just seen for the electron and proton in a hydrogen atom, or it can mean a change of direction without a change in speed. This latter characteristic of acceleration is familiar from the experience of going around corners in a car: you are pushed against the door or the side of the seat and experience a force that turns you with the car. The force is larger the faster you are moving as you enter the turn, and this is an indication that you are accelerating, even though you may not be speeding up or slowing down.

Even this kind of acceleration, when experienced by charged particles, causes them to radiate light. Imagine a bunch of electrons forced to move in a circle, as if they were stuck on the rim of a rotating wheel. They generate light, the wavelength of which gets shorter and shorter (so the photon energy gets larger and larger) the faster they move around the circuit, because of this angular acceleration. The light generated in this way is called *synchrotron radiation*, and it is a common means to generate X-rays. Particle acceleration is also related to the light seen in the northern and southern *aurorae*, which are produced when charged particles from the Sun are forced to spiral by the Earth's magnetic field as they enter the atmosphere where they hit molecules and cause them to emit light, as described in the next section.

Quantum light emission processes

These basic mechanisms underpin all light sources. But the details of how the atoms act as a group can have a strong influence on the characteristics of the light that is eventually emitted—as I noted in Chapter 1, a light bulb emits a very different sort of light than a laser pointer, for instance. In order to understand this, we need to delve a little deeper into the structure of atoms, since the process of emitting light from an atom is not completely encompassed by the analogy with swings. Something needs to be added in order to accommodate the fact that atoms and molecules are quantum mechanical entities.

For our purpose this simply means that the electrons in the atom or molecule can only hold energy in fixed amounts. Using our swing model, it means that the maximum extension of the swing amplitude cannot be anything you like. Rather, it is restricted to certain particular values: it is quantized. More particularly, the energy of the swing comes in discrete packets, or quanta, and when you push it you can only make it jump by one or more quanta. In the atom, this means the energy of the electron can change only in the same discrete units when it absorbs or emits a single photon. The energies involved in making these jumps are very small by everyday standards. Take a light in your house. This consumes energy at a rate of, say, 60 W, or 60 Joules per second. A single photon emitted by the atoms in the light bulb possesses about one billion billionth of a Joule. So a light bulb is emitting more than ten billion billion photons per second.

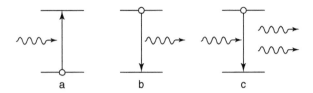

26. An atom undergoing a. absorption, b. spontaneous emission, and c. stimulated emission.

An atom can be prepared in an excited state simply by shining light of the right frequency on it, as shown in Figure 26a. (Of course, this begs the question, but we've also seen that the excitation can also be accomplished in other ways, for instance by running an electrical current through the medium.) Now, according to quantum theory a corollary of the 'jumpiness' of the electron motion is that the atomic electrons are pretty much stable against the emission of light once in one of these discrete states. They are like a ball on a shelf in a cupboard—in principle it can lower its energy by dropping to a lower shelf, but this cannot happen in practice unless you give it a little push so that it rolls off the shelf.

So it would appear that quantum physics suggests that atoms cannot emit light, since once you have put them in these stable states—that's it. Well, it turns out that, except for the lowest-energy state of the electron in the atom, a push is available to allow the electron to drop from a more energetic state to a less energetic state. And the surprising thing is that the shove arises from nothing.

In Chapter 4, I noted one of the strangest features of quantum physics: even the emptiness of space is a seething background of activity, filled with 'vacuum fluctuations'. These fluctuations in the electromagnetic field can cause the atomic electron to drop to a lower energy level and to give up the energy difference in the form of emitted light. This process, by which an atom (or molecule) undergoes a transition from one stable state (the excited state) to another of lower energy (the ground state), emitting a photon on the way, is called *spontaneous emission* (Figure 26b). It is something that each atom does on its own. It was used by Einstein in order to account for the proper energy balance between a light beam and the matter on which it was shining. If spontaneous emission does not occur, then the atoms hold on to the energy from the light beam and the situation we see everywhere around us, where most things are in a stable state in equilibrium with their surroundings, would be impossible.

Einstein understood the central mystery of spontaneous emission: that it was a random process. You just cannot say exactly when any given atom will make a jump. All you can say is that on average, after some amount of time (that depends on the particular atoms, but is typically less than a microsecond), in a large collection of atoms, about two-thirds of them will have emitted a photon. But the origin of the fundamental randomness remained a mystery until, in 1927, Paul Dirac's quantum field theory introduced the idea that quantum vacuum fluctuations were at the root of this. The notion that a field containing no photons at all can cause an excited atom to be unstable is at odds with our intuition, and it took until the 1950s for Lamb's measurements to confirm that Dirac's view of the vacuum was right.

What this means is that even everyday occurrences—the picture generated on a TV screen by means of LEDs, for example—have at their heart this fundamentally random characteristic arising from quantum mechanics. By contrast, the emission of light by atoms when they are pushed to give up their energy by the application of another light field is called *stimulated emission* (Figure 26b). This form of recouping energy from atoms into the light field does not have random character. And this makes possible a very different kind of light: that of the laser.

Coherence: acting all together

When atoms and charged particles behave individually, 'doing their own thing', the light they emit when there are many of them is a sort of uncoordinated set of waves. Even a very small LED, with a size of much less than a millimetre, contains a vast number of atoms, so this is a common situation.

A feature of this uncoordinated emission is that each atom emits its photons at random, with no acknowledgement of what the adjacent atoms are doing: the light goes off in many different directions, and the photons are all emitted at different times. In effect, the randomness of the emission process is reflected in the randomness of the resulting light intensity.

Let's say we put a photo detector in front of a light bulb. (A photo detector works like a light bulb in reverse. It uses the photoelectric effect—light incident on it produces an electrical current that can be measured.) What we would see is that the electrical current from the detector was very noisy, because the light incident on the detector has an electric field that changes quickly and randomly, corresponding to the arrival of random numbers of photons at each instant of time.

But what if it were possible to coordinate the atoms, so that they acted together? We can think back to our earlier analogy: imagine a collection of swings, each with the same oscillation frequency. The swings may be all oscillating at random: that is, with each at a different point in its repetitive trajectory at any instant. Or, they may be in synchrony, with the differences in trajectory between adjacent swings being a fixed amount—like a wave caused by the adjacent spectators in the crowd at a football match standing up and sitting down in sequence. In the first case, the light that is emitted from these uncorrelated atoms is like that from a light bulb or LED and is said to be *incoherent*. In the second case, however, the atoms oscillate in lock-step, and the light they emit is given off in a *coherent* fashion—all of the photons are emitted in the same direction. This is what happens in *stimulated emission* (Figure 26c), and is the basis for the laser.

Laser light

The laser is perhaps the most important invention in optics in the last century. This device produces extraordinarily useful beams of light that have revolutionized the range and capabilities of applications. Not only is it a specific source of illumination, in microscopy and spectroscopy for example, but it also provides a means to direct significant energy on to a particular target in a tailored way, and thereby to control the dynamics of matter. An extreme example of the latter application is laser-driven fusion of atoms, discussed in Chapter 7, that may enable new forms of nuclear energy which can draw upon a very large supply of fuel.

A laser consists of two elements. The first is an optical amplifier, or gain medium, that generates light from atoms by means of stimulated emission. The atoms are placed between two mirrors (and possibly other optical elements) that form an *optical cavity*. The number of photons in the cavity builds up as the atoms emit light until there is a balance between the energy put into the light beam from the atoms, and the energy leaking out of the cavity through the mirrors. As the amplifier is turned on, by providing a means to excite the atoms, the light emitted from the amplifier is reflected back into it by the end mirror of the cavity. That causes further stimulation of radiation from the excited atoms, and thus the light in the cavity increases in brightness. At the other mirror, some of the light is transmitted out of the cavity as a useful output. Some is reflected back into the gain medium. When the rate at which light is put into the cavity by the amplifier equals the rate at which it is extracted through the output mirror, the laser is said to be at threshold. Beyond this point, any increase in the amplifier gain (the rate at which atoms are put into their excited states) leads to an increase in the intra-cavity intensity, and thus to an increase in the output light.

The optical cavity also imposes a restriction on the colours of the laser. It turns out that the frequencies that experience the most gain are those for which the light waves add with constructive interference on each round trip. This means that the length of one round trip of the cavity should be equal to an integer multiple of the wavelength. The frequencies that satisfy this resonant condition are said to be the *cavity mode* frequencies.

The reason that lasers are important is related to the fact that the light they emit is coherent: all the photons go in more or less the same direction, with the same colour. The direction is defined by the laser cavity, and the colour by the atoms in the gain medium and the allowed modes of the cavity. This leads to the property that the light is in the form of a beam—the laser beam—which is about as close to a 'ray' of light as you can imagine. It diverges as it propagates due to diffraction, but has the smallest divergence possible. This property also means that it can be focused to a very small spot using lenses or mirrors.

A second feature that contrasts the light emitted from lasers with that from light bulbs is that laser light is usually very pure in colour. In other words, it consists of only a narrow range of wavelengths, whereas lamps often emit a broad range of wavelengths. The light intensity is very stable (registering as very low noise in a photodetector output), and the light can be emitted continuously or as a sequence of pulses.

The ability of laser light to be focused to a very small spot makes it useful for microscopy, and there are a number of different ways in which, by scanning the laser spot across an object at the focus of the microscope lens, and detecting the light scattered or re-emitted from the object, three-dimensional images of the object can be made. This approach is very useful in looking at animal tissue, for example, and optical microscopies of this kind are widely used in biomedicine.

Many applications of lasers in manufacturing also stem from this property of laser light. The ability to mark, cut, drill, or weld metals, for instance, requires that a lot of energy be deposited in a small region of the metal in a short time. So high-power lasers producing coherent light beams in the form of pulses that can be focused are ideal for such materials processing.

Similar properties are needed for some medical applications of lasers, also involving materials—this time skin, teeth, or hair. Laser correction of vision and laser dentistry are now commonplace, as is the use of lasers to remove tattoos by heating up the ink until the drops break up, and for hair removal— although unfortunately not hair regrowth! Other familiar devices that make use of laser light's ability to address a very small spot are the CD, DVD, Blu-ray, and some computer disk storage devices. Very tiny spots of light in the recording medium allow the very high-density storage of data in the material.

The very pure colours achievable with laser light make it possible to distinguish the constituent atoms and molecules of different mixtures by means of spectroscopy. As noted in Chapter 1, different atoms, and indeed different molecules, have characteristic frequencies at which they absorb and radiate light, due to their different structures. Extending the analogy developed in this chapter, they are like swings in which the chains or ropes holding the seat are of different lengths—their natural oscillation frequencies are dependent upon the way in which they are put together.

In fact, each atom and molecule has a range of different absorption and emission frequencies, corresponding to the excitation of different electron configurations. Typically these lie in the blue part of the visible light spectrum, but some molecules absorb at much shorter wavelengths, invisible to humans. Many molecules also absorb light at wavelengths longer than the red end of the visible spectrum. This absorption arises from the vibrations between the atomic nuclei that make up the molecule. Since nuclei are much heavier than electrons, they tend to oscillate at much lower frequencies. This set of frequencies is a sort of molecular 'fingerprint' that enables identification of a particular species.

The catalogue of these fingerprints is of importance, of course, in chemistry, since it allows the different elements involved in a reaction to be identified. It is also used in molecular biology, and even in cell biology, when the movement of particular 'tag' molecules can be studied. It is critical, too, for astrophysics, in which the elements present in astronomical objects—stars, galaxies, nebulae—can be determined, as well as in atmospheric physics and meteorology, for the remote sensing of pollutants or particles. Such monitoring provides key data in assessing the impact and origin of climate change.

By combining the outputs of several different lasers—say one emitting red light, one green, and one blue—it is possible to make a laser projector. By changing the intensities of each laser individually according to the video signal output of a computer or Internet link, perhaps by means of a liquid crystal cell, movies can be projected on a big screen with vivid, highly saturated colours. The combination of red, green, and blue (RGB) light is sufficient to make up a complete colour palette and lasers produce very bright images on a screen.

X-rays

When the wavelength of the light is very short, in the X-ray region of the spectrum, a different kind of spectroscopy arises. X-ray photons are energetic enough to excite the most tightly bound of the electrons in atoms—not just the outermost electrons. This means that X-rays can be used to look into the heart of atoms and molecules, and to understand their local environment, which can shift the binding energies of the electrons. X-ray absorption spectroscopy is widely used in the study of materials for a variety of applications, from detecting trace pollutants to understanding the structure of glasses. As noted in Chapter 3, X-rays are also used to study the structure of crystals by means of diffraction. When the X-ray wavelength is close to the spacing between atoms in the crystal, then the crystal acts as a 'diffraction grating' and scatters the X-rays in discrete directions. By detecting these diffraction patterns on a camera, it is possible to reconstruct the three-dimensional structure of highly complex crystals using advanced inversion algorithms. Today this is a routine process for characterizing isolated biologically and chemically relevant molecules, determining the structures of possible new molecules in order to design them for a specific function.

Some of the best light sources for this sort of spectroscopy are synchrotrons. In order to produce X-rays at the required short wavelengths, synchrotrons have to produce very energetic electron beams, and accelerate them around a big ring. Experimental stations catch a glint of light as the electrons rush by, leading to short bursts of X-rays that are used for diffraction imaging. For example, the Diamond Light Source at Harwell in England accelerates electrons to more than a billion volts in a ring that is more than half a kilometre long. The next generation of X-ray light sources is being built using linear particle accelerators, which produce extremely bright X-rays beams. The X-ray diffraction pattern shown in Figure 20 was taken using the Diamond synchrotron.

Ultrashort light pulses

Laser light can also come in short bursts. There are several ways to arrange for this to happen. The method that produces the briefest pulses is called *mode-locking*. This requires a gain medium that has a large bandwidth—that is, it can amplify light over a broad spectrum. This allows several of the modes of the optical cavity to experience gain. If it is also arranged that these modes all have the same phase, then light waves of many different frequencies add constructively to yield a single pulse inside the cavity, bouncing back and forth between the mirrors. The brevity of the pulse is determined by how many frequencies are locked—the wider the range of frequencies, the shorter the pulse.

The possibility of creating very short duration laser light pulses enables a kind of measurement called dynamical or time-resolved spectroscopy. It allows us to see how things change in time, based on an old principle: the stroboscope. The general idea of how you use light to 'freeze' rapid motion stems back to the work of Eadweard Muybridge in the late 19th century. He invented the idea of using a fast camera shutter to photograph a horse trotting. The legs of a horse move too fast for the human eye to resolve, and it was not known at that time whether all four legs left the ground at any point during the stride. To settle the matter, Muybridge set up a bank of cameras, each of which had its shutter opened by a trip wire, which was triggered when the horse passed the camera. This enabled him to 'slice' a short pulse from the light reflected from the horse. The duration of this pulse of light was shorter than the time over which the horse's legs moved. The outcome of his work was that he was able to inform Leland Stanford, the funder of his research, that in fact there is a point in a horse's stride at which none of its legs touch the ground.

Mechanical shutters on conventional cameras could close very fast, but not fast enough to see some forms of animal motion, such as the flapping of the wings of a hummingbird. Even faster physical events such as an explosion, where things change on timescales of thousandths of a second, were out of reach. To solve this problem, Harold Edgerton at MIT in the 1950s invented a new kind of non-mechanical shutter, based on an optical switch. He was able to take 'still' photographs of explosions, for instance, using this device.

These shutters are what we might call 'passive' instruments. They simply allow a slice of light to pass through when they are open, so are suitable to events that are well illuminated (the horse in California sunshine) or emit a lot of light themselves (the explosion). But we can imagine an 'active' instrument, one that generates a short pulse of light to illuminate a moving object. Think of the light pulses emitted by the flash unit on a camera. A flash of light with a duration that is short compared to the time taken for the thing we're looking at to move provides an image of the object 'frozen' in time, even when using a camera shutter speed that is much longer than the changes of the object. A second flash freezes the motion at a later time. The same is the case for subsequent flashes.

A movie composed of the sequence of frames taken on repeated trials of the event reveals very rapid changes in the system, on a

27. A bullet frozen in motion using stroboscopic imaging.

timescale that is much faster than could be observed by the eye. Indeed, the brevity of events that can be seen in this way is truly breathtaking. Edgerton invented a 'stroboscope' of this kind in 1931, and some of his most iconic images, such as a bullet passing through an apple or a playing card (Figure 27), were taken with this device.

Using modern pulsed lasers as the 'flash' it is possible to observe not just the frozen motion of bullets but even the motion of atoms in a molecule involved in a chemical reaction (for which the 1999 Nobel Prize in Chemistry was given to Ahmed Zewail) and the much, much faster motion of electrons whizzing around the nucleus of an atom. The timescales for these motions are staggeringly small—less than one tenth of a million-millionth of a second (100×10^{-15} seconds, or 100 femtoseconds (100 fs)) in the case of molecules, and a few tens of billion-billionths of a second (100×10^{-18} seconds, or 100 attoseconds (100 as)) in the case of atomic electrons. These fields—femtochemistry and attoscience respectively—are at the forefront of what is possible in the interaction of light and matter. I shall consider them further in Chapter 7.

Chapter 6
Light, space, and time

Robert Grosseteste, Bishop of Lincoln and the first Chancellor of the University of Oxford in the 13th century, was one of the leading thinkers of his day, and a proponent of the works of the ancient Greeks. For him, as for many philosophers, the challenge of understanding light's nature was critical to understanding the world. In Grosseteste's treatise on the subject, entitled *De Luce*, he extols the primary importance of light: 'The first corporeal form…is in my opinion light. For light of its very nature diffuses itself in every direction in such a way that a point of light will produce instantaneously a sphere of light of any size whatsoever'.

For Grosseteste, light defines space by its propagation instantly throughout the universe. Without light, there is no space, and therefore no forum in which events can take place. Matter, and thus the spatial extension of objects, are coupled to light, but cannot be separately defined. This intimate connection between light, space, and matter—in Grosseteste's hands amenable to quantifiable description—informed the development of ideas regarding cosmology in the subsequent centuries.

Space-time

For Newton, space neither admitted nor demanded definition. He thought of space as a pre-existing entity, a sort of theatre in which events played themselves out. Large-scale motion of bodies in the heavens was integral to his idea of a set of universal laws. Einstein, by contrast, places light at the centre of space. For him, it defines space and time by virtue of setting the speed limit for signals sent from one part of the universe to another. The fact that there is a finite maximum speed turns out to make space and time inseparable. Einstein's theory of relativity teaches us that we cannot think of one without thinking of the other. This is because our perception of space and time is based on local measurements of distances and time intervals. These measures appear differently to those moving relative to us, because of the speed limit imposed by light.

How does this strange entwining of space and time by light arise? Let's start with Newton's conception of space. We can think of this as a sort of scaffold—a collection of imaginary rods of fixed length all connected together in a three-dimensional framework, as shown in Figure 28. Newton thought that this sort of structure pre-existed any event, and indeed that all events took place somewhere in this structure. Events can therefore be specified by their distance from a fixed point in the frame by counting the number of rods to reach the location of the event. Of course events occur at a certain time, too, so the scaffold must be equipped with clocks to measure the time. If a clock is placed at the junction of each of the rods, they all show the same time everywhere in space, and we can easily define a 'universal time'.

Now we must ask several questions. First, how should we build a clock? Second, how should we ensure that they are all synchronized across space? Third, how should we build a ruler? These questions all have answers that are intimately related to the properties of light. Indeed, the answer to the last is this: one metre is the length of the path travelled by light in a vacuum during a time interval of $1/299,792,458$ of a second. It is therefore linked to the answer to the first question: how accurate a clock can be built.

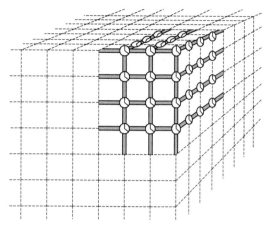

28. A scaffold representing space made up of a 3D lattice of measuring rods. At each intersection is a clock, all synchronized.

Clocks

The primary characteristic of a clock is that it 'ticks'. That is, it signals at regular intervals of time. By counting the number of ticks between events, we can assign a time interval between them. The more precise a clock is, the more regular the intervals between the ticks. In a grandfather clock, the 'ticks' are provided by the regular swings of a long pendulum. In an electronic wristwatch, the 'ticks' are the oscillations of a piece of quartz crystal. These are much more regular than the swings of a pendulum, which can be affected by the temperature and humidity of the place where the grandfather clock is located. Therefore the quartz watch is said to keep better time than the grandfather clock.

The most accurate clocks in the world are based on the highly regular clicks of electrons moving around in atoms. We've seen that atomic electrons can jump between different stable energy levels within an atom. For some atoms, the difference in energy between these levels is extremely well defined, and the rate at which electrons jump between the two can therefore be used to

85

define a set of ticks, based on the frequency of the light used to push the electrons up and down between the two energy states.

The idea is simple enough: illuminate an atom with microwaves (recall that microwaves are just like light, but with a much lower frequency—billions of oscillations of the electric field per second rather than millions of billions as in the case of visible light). Then slowly change the frequency of the microwaves until the electrons move between the atomic energy levels most efficiently. This *defines* the number of cycles per second of the microwave radiation—the rate of ticks of the clock—in terms of the energy level spacing of electrons in a particular atom.

There are many technical challenges to building such an atomic clock—cooling the atoms, preparing them in the right initial state, illuminating with microwaves in a clever way to maximize the sensitivity, detecting that the electrons are indeed in the higher state at a given time—but clocks based on caesium atoms are now the most accurate way to measure time, with a rate of ticking that deviates from complete and utter regularity by only one second in 300 million years.

These clocks provide a time standard that is agreed internationally and maintained by government laboratories such as the National Institute of Standards and Technology (NIST) in the US, the National Physical Laboratory (NPL) in the UK, and Physikalisch-Technische Bundesanstalt (PTB) in Germany. They are crucial components of many technologies that underpin our daily lives. For instance, they are vital for the global positioning system (GPS) that is the basis for navigation, including the satnav commonly used in cars.

Clock synchronization

The next challenge is to synchronize the two clocks, so that they are calibrated. One way to do this is by sending a signal from one clock to another. You start the first clock, sending a pulse of light to the other clock telling it when your first 'tick' occurred. The person in charge of that clock then knows what fraction of tick her clock is behind yours (since the clocks are the same construction, we can assume they tick at the same rate) and can use this information in setting the correct time.

There's an interesting consequence to this. Imagine you want to synchronize your clock on Earth to that on a distant planet, in a far-away galaxy. You send your pulse of light off into the direction of the planet, and then you wait. As the planet is far away, it might take a very long time for the light to get there, even given the high velocity of light. Meanwhile you are getting older and older. The person who receives your synchronization message will have received it from the young you—she will see you as you were when you sent the message.

Likewise when we look at the night sky, and see the distant stars, we are seeing images of them by receiving light that left their surfaces a long, long time ago. And as we look to stars or galaxies that are even farther away, so we look into the deeper and deeper past, seeing the universe as it was billions of years ago. In that sense, the light that reaches us is also billions of years old—it has been travelling across space for an age since its birth in the remote past. Light is the oldest thing we can see in the universe.

Our clocks, though, are a bit closer together. An interesting fact is that when you put one of these clocks on an aeroplane moving at 800 kph or so, you find that it ticks slower than one on the ground. That is, if you set the clocks to tell the same time, you will find that the one in the aeroplane appears from the ground to be ticking at a slower rate than the ground-based one. This is a consequence of the maximum speed at which signals can be conveyed between the two clocks—the speed of light.

You can see this by looking at the situation in Figure 29. There, person A is on the ground and person B is in an aeroplane moving

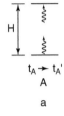   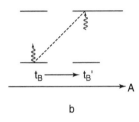

a   b

29. Time dilation due to relative motion. Observers A and B both measure the time for a pulse to arrive at the suspended mirror. They measure different times because they are moving relative to one another.

at high speed. B watches as A sends a light pulse to a mirror suspended at height H above the ground. From A's point of view, the light pulse travels distance H. However, from B's perspective that signal will travel a slightly longer distance than H since A appears to be moving backwards at high speed relative to B. Therefore, because the signal travels at light speed according to both A and B, and both record the same number of ticks between sending and receiving the signal, then the only explanation is that B's clock is ticking slower than A's clock when seen from A, and A's clock is ticking slower than B's when seen from B. This effect is called *time dilation*.

Einstein argued in a similar way that space should also contract. That is, a rod of B's imaginary scaffold should look smaller to A than an identical rod in his possession. And vice versa: A's rod looks shorter to B than her own.

Both of these effects arise because there is a maximum speed at which signals of any kind can propagate, and this speed is the same for everyone. If that were not the case, then one could determine a preferred scaffold, or 'frame of reference', in which the signals went at the highest speed. Einstein's work on relativity showed that there is no preferred frame, so that Newton's idea of a fixed, pre-defined space could not be the case. Since the maximum speed for signals of any kind turns out to be light speed, light is crucial in defining space and time.

It's worth asking how precisely we might be able to synchronize two clocks by the 'light pulse' method discussed previously. You can see that one way to do it is by making the light pulse as brief as possible, so that the uncertainty over when it arrives is minimal. Thus it is important to know if there are limits to the brevity of light pulses. It turns out that there are, and they arise from a similar sort of wave property that limits the resolution of imaging systems, as we saw in Chapter 3.

We might start by asking how it is that we determine the frequency of a light wave. Imagine taking one of our clocks and asking how many peaks of the wave reach us during the interval between two ticks. The greater the number of peaks, the higher the frequency. The precision with which we can determine the frequency depends on how many times we repeat this measurement, since our ability actually to tell whether we have reached the peak of a wave may not be perfect. Thus, the longer we count, the more precise our determination of frequency. This trade-off is fundamental for waves—the imprecision in frequency multiplied by the uncertainty in the time interval is a fixed product. This was first understood by Joseph Fourier, a 19th-century French mathematician and scientist who played a key role in formulating the wave model of light propagation.

Ultrashort light pulses

Fourier's theorem is important for clock synchronization, since it says that if we want a short pulse, we must have an indefinite frequency. Another way to say this is that a short burst of light is constituted by a broad range of colours. That's entirely analogous to the situation Abbe identified for imaging optics: a high-resolution image, demanding a small focus, requires a wide range of ray angles to be collected. In fact, the analogy goes further. Just as Abbe showed the smallest size of a focal spot could be approximately one wavelength of the illuminating light, so Fourier showed that the shortest-duration pulse is a single cycle of the field.

What this means in practice is that for light in the visible region of the spectrum, it is possible to produce pulses with a duration of about 2 fs. Amazingly, light sources derived from lasers can now routinely produce pulses of such startling brevity. They are based on the mode-locking technique described in Chapter 5.

But these are not the shortest bursts of light that occur in nature, nor even the shortest that can be produced in a laboratory. That honour goes to light sources with much shorter average wavelengths. For instance, using the single-cycle argument you can easily see that if the wavelength is shortened, then the duration of an optical cycle is reduced and in principle the duration of the pulse can be reduced. This approach currently holds the record for the world's shortest controllable light pulses. By shining very intense laser light on an atomic gas, a process known as high-harmonic generation produces light waves with a frequency multiples of several tens of the driving laser frequency. This allows pulses with durations of several tens of attoseconds (10^{-18} s, or a billion billionth of a second). These are unimaginably short bursts of light, with duration equal to the time it takes an electron to oscillate within an atom.

Frequency combs

In Chapter 5, I stated that in a mode-locked laser a single pulse traverses the optical cavity. Each time it bounces off one of the mirrors, a little bit of the light pulse is transmitted through the mirror and exits the cavity. As a consequence, outside the cavity the light appears as a 'continuous' sequence of pulses spaced by the round-trip time of the light in the cavity, typically a billionth of a second or so. A 'snapshot' of such a train of pulses would show these very short bursts separated from each other by a delay that is long compared to their duration, like the teeth of a comb (as shown in Figure 30). And the pulses can be made identical to one another by careful adjustment of the laser producing them, so that the electric field of each of the pulses peaks at exactly the same time with respect to the intensity envelope of the pulse.

It turns out that this configuration means that each of the 'teeth' in the frequency comb has a very precise position at an absolute frequency. A precisely calibrated set of frequencies is a very important tool for building accurate clocks. This is because it allows a direct comparison of optical frequencies to lower (usually microwave) frequencies, which can be counted by means of electronics.

Thus frequencies in the microwave region inhabited by the caesium atomic clock can be compared simply to much more precise electronic transitions in the optical region of the spectrum in, for example, strontium atoms or aluminium ions. Therefore the standard caesium atomic clocks used in satellite navigation,

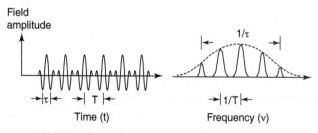

30. **Train of identical nearly single-cycle optical pulses. The spectrum of the pulse train looks like the teeth of a comb, hence it is called a frequency comb.**

for instance, can now all be synchronized and to tick at the same rate to within one part in a thousand billion (i.e. $1:10^{-12}$), due to the precision of the optical electron oscillation frequency.

'Optical clockwork' of this kind allows the comparison of disparate frequencies with such remarkable precision that it provides a means to test the tenets of relativity, and thus to understand better the role of light in defining space and time. Frequency, and thus time, is the physical quantity that can be measured with the highest precision of any quantity, by far.

Optical telecommunications

Frequency combs are also important in telecommunications links based on light. In Chapter 3, I described how optical waves could be guided along a fibre (or alternatively in a waveguide inside a photonic integrated circuit on a glass "chip"). This phenomenon underpins the long-distance telecommunications infrastructure that connects people across different continents and powers the Internet. The reason it is so effective is that light-based communications have much more capacity for carrying information than do electrical wires, or even microwave cellular networks. This makes possible massive data transmission, such as that needed to deliver video on demand over the Internet.

Many telecommunications companies offer 'fibre optic broadband' deals. A key feature of these packages is the high speed—up to 100 megabytes per second (MBps)—at which data may be received and transmitted. A byte is a number of bits, each of which is a 1 or a 0. Information is sent over fibres as a sequence of 'bits', which are decoded by your computer or mobile phone into intelligible video, audio, or text messages. In optical communications, the bits are represented by the intensity of the light beam—typically low intensity is a 0 and higher intensity a 1. The more of these that arrive per second, the faster the communication rate. The MBps speed of the package specifies how rapidly we can transmit and receive information over that company's link.

Why is optics so good for communications? There are two reasons. First, light beams don't easily influence each other, so that a single fibre can support many light pulses (usually of different colours) simultaneously without the messages getting scrambled up. The reason for this is that the glass of which the fibre is made does not absorb light (or only absorbs it in tiny amounts), and so does not heat up and disrupt other pulse trains. This is very different from electrons moving down a copper wire, which is the usual way in which local 'wired' communications links function. Electrons tend to heat up the wire, dissipating their energy. This makes the signals harder to receive, and thus the number of different signal channels has to be kept small enough to avoid this problem.

Further, a light beam propagating in glass has to be very intense in order for it to influence another light beam. For instance, when you cross the beams from two laser pointers you don't see either beam distort or deviate from its original path even though the two beams pass right through each other. (If your laser pointers had enormous power, you might just see such an effect, but only because the room is full of air. In a vacuum they would still not influence each other.) This means that the 'crosstalk' between light beams is very weak in most materials, so that many beams can be present at once without causing a degradation of the signal.

Second, light waves oscillate at very high frequencies, and this allows very short pulses to be generated, as described earlier. This means that the pulses can be spaced very close together in time, making the transmission of more bits of information per second possible. Indeed rates of 40 Gbps (a Gb, or Gigabit, is a billion bits) are possible in current-generation commercial long-haul systems. Electrical signals in copper wires are limited in the duration and spacing of pulses coding the information by the heating and other effects noted previously, which tend to get worse at higher frequencies. Copper wires run out of steam, as it were, at much lower bit rates.

Fibre-based optical networks can also support a very wide range of colours of light. Glass transmits a broad range of wavelengths, with particularly low scatter and absorption loss in a spectral window from 1.3–1.55 μm. The rate at which photons are lost in fibre at these wavelengths is about 5 per cent per kilometre. These losses can be made up by amplifying the light while in the fibre so that transmission over very long distances (such as across the Atlantic Ocean) is possible without any conversion of the light to electrical signals or vice versa.

The telecommunications spectral window is divided into many individual frequency 'slots'—much like the frequency comb shown in Figure 30. Each spectral component is a separate communications channel. There can be 150 or so slots in the window. Within each channel, a 40 Gbps optical signal can be operated. Therefore the total bit rate for communications is 150 times 40 Gbps, or 6 Tbps (1 Tb, 1 terabit, is 1,000 billion bits).

The ever-increasing demand for communications bandwidth due to increased use of the Internet and the services it provides have spurred optical engineers to new heights of creativity in harnessing the potential of light.

Chapter 7
Lighting the frontiers

Despite its long history as possibly the oldest continuous branch of natural philosophy and science, optics remains at the forefront of research and application. It is ubiquitous: as a tool for sensing, imaging, and communications, as well as providing ways to explore, discover, and illustrate new fundamental effects.

Light can generate conditions at the extremes of what is known to be possible according to physics, such as extremes of temperature and extremes of pressure and stress that do not exist naturally, except perhaps in the remotest of stars. And light can be used to observe and even control really fast events, such as the movement of electrons inside atoms.

Further, light can exhibit strange features associated with the quantum world, revealing even in everyday conditions some of the counter-intuitive aspects of the fitful world of randomness that underpins the stable, solid world of our normal experience. In this chapter, I will explore some of the frontiers to which, and across which, light is currently taking us. Exploration of these frontiers is possible because of the great technological strides that have been made in light sources, optical systems, and detectors, which enable exquisite control over the shape and intensity, in both space and time, a light beam can take.

Light mechanics

Light can exert forces on objects. This allows 'remote control' of bits of material using shaped light beams. Light can be used to move matter around, and bring it into contact with other objects, or to manipulate the internal configuration of atoms and molecules, forcing them into, for example, simple chemical reactions, in ways that allow both the study and exploitation of unusual material properties. That's an extraordinarily powerful feature in many areas of research and study.

The concept of mechanical force arising from light has its origins in the momentum carried by each photon. For instance, when a photon is reflected from a mirror, that mirror experiences a force that provides the exertion needed to redirect the photon, just as water from a fire hose hitting a wall exerts a force on the wall by virtue of its bouncing off.

Similarly, when a photon is refracted it changes direction, and this, too, requires a force. Thus the photon exerts a force on the refracting element. If a light beam is incident on a glass bead, rays that are nearly tangential to the bead will experience the greatest change in direction. A photon traversing the ray in the lower half of the bead is directed upward as it propagates through the glass surfaces. The bead therefore experiences a force in the opposite direction. Since the momentum of the photon in the forward direction (the direction it was moving before encountering the bead) is reduced, there is also a net force in the forward direction. The strength of this force depends on how many photons are refracted per second. A light beam that is more intense at a position near the centre of the beam than on its periphery will therefore drag the bead towards the higher-intensity part of the beam.

This effect can be used to make a focused light beam into an 'optical tweezer', which is able to hold on to a minute object and move it around as the light beam is steered. Optical tweezers find applications, for example, in biology, by enabling control of the position and movement of individual strands of DNA and the characterization of tiny molecular motors. Specifically, DNA, proteins, and other biologically important molecules can be stuck on to these beads and can therefore be handled using optical tweezers. Their position can be controlled with a precision of much less than the wavelength of light. This allows extremely small forces to be measured, such as happens when biological cells adhere to surfaces or other cells. It also enables manipulation of objects, such as holding cells in places as they are operated on using other lasers (so-called cellular surgery). Optical tweezers can be combined with various other test methods, such as light scattering from aerosols, or spectroscopy, to characterize particles that may be pollutants in the atmosphere.

These opto-mechanical forces can also be used to access completely new states of motion of small objects. In particular, it

31. A nano-scale cantilever controlled using light forces. The discs are tiny mirrors about 30 μ m in diameter.

is now possible to build tiny mechanical cantilevers, illustrated in Figure 31, and to both observe and control their motion using light. Light forces can be used to cool or heat the oscillations of the cantilever—like running down or winding up a watch spring—and eventually to bring it close to the quietest state possible, where only quantum fluctuations of the motion disturb the complete stasis of the lever. Light forces can also be used to cool atoms—much smaller objects—and this reveals even more strange quantum states of matter.

Ultra-cold

What's the coldest thing you've experienced? Colder than winter in Oxford (approximately 2°C), or in Ottawa (−20°C), or the South Pole (nearly −90°C)? Or perhaps the effects of liquid nitrogen, at −200°C. These are certainly cold, but by no means the coldest things possible. It turns out that there is a lower limit for temperature: −273°C, or 0 K (Kelvin), below which it is not possible to cool things further. This is the temperature at which things are as still as they're going to get, with just the effects of quantum mechanics to cause a little jiggling about of atoms and molecules.

It's not actually possible to build a machine to get to absolute zero but it is possible to get very close using an 'optical refrigerator'. In fact, you can get cold enough to make the atoms almost stop moving. What this means is that their size gets bigger. (Quantum mechanics tells us that you can't simultaneously specify the precise location and speed of an object. If the atom is completely stopped, then it must be extended over all space.) Therefore all atoms in the cloud that has been refrigerated occupy the same region of space, and this gives rise to some very strange new phenomena.

An optical refrigerator works by using lasers to 'cool' atoms. Imagine a laser beam shining on an atom that is moving from left to right, say. The laser shines from the right to left, so that a stream of photons hits the atom. The laser is tuned in frequency to be absorbed by atoms that are moving at a particular velocity. Now, when the atom absorbs a photon from the laser beam, it gets a kick from the photon, and thus slows down. (More specifically, the momentum of the photon is transferred to the atom. Since it is in the opposite direction to the initial momentum of the atom, it reduces the momentum, and thus the speed of the atom.) The atom must re-emit the photon at some later time, and it will get a kick in the opposite direction to that in which it emits the photon. But the direction in which it re-emits the photon is random—it can go in any direction at all.

If you look at enough of these absorption-scattering events, then you will find that, although the light is always absorbed from one direction (the incoming laser beam) it is emitted uniformly in all directions—no one direction is preferred. The consequence of this is that on average a collection of atoms moving in a direction opposite to the laser beam grinds to a halt, and is left with random motion representing a temperature that is inversely proportional to how long it holds on to the light before re-emitting it.

There are several refinements of this approach, each of which uses light to cool atoms (and molecules) to even lower temperatures, and for which light acts like a 'viscous fluid' in which the atoms move slower and slower. It is even possible to use light to trap atoms using optical tweezers once they are slow enough. This allows the application of yet more sophisticated optical cooling techniques, by which it is possible to get to temperatures of a billionth of a degree above absolute zero.

I referred to some residual 'jiggling' of the atoms that happens even at zero temperature, arising from quantum mechanics. The range of this jiggling can be thought of as the spatial extent of the atom itself. That is, according to quantum mechanics the atom isn't just wandering around in a random fashion over a small region of space, but rather it is actually present across all that region at once. For atoms trapped at such low temperatures the size of that region may be several thousandths of a metre. That's a remarkably large atom, given that the distance of the electron from the atomic nucleus is less than one tenth of a billionth of a metre. What's even stranger is that several atoms can occupy this region of space at the same time.

That's conceptually very counter-intuitive. We often think of atoms as being like little billiard balls, that can be packed close together, as in a solid material, but which retain their individual distinction by virtue of their location inside the material. That's not the case for these very cold atoms. They are each everywhere at once, in a new state of matter identified by Einstein based on the discoveries of the Indian scientist Satyendra Nath Bose and called, not surprisingly, a Bose–Einstein condensate.

This very strange state has some remarkable properties. For instance, it is a superfluid, which flows without viscosity. Further, it is possible to split the entire atomic cloud in half and recombine it to show quantum interference between the two separated parts, essentially demonstrating that a big object (containing many atoms and of a palpably visible size) exhibits quantum character, attributable to the uncertainty of whether an atom is in one part of the cloud or the other. One has to think of each atom occupying both separate components at the same time.

Because these cold atoms can be trapped in light beams, it is also possible to create spatial structures out of several light beams that can be used to manipulate the atoms. For instance, when two light beams coincide they form an interference pattern (see Chapter 3) in which there are regions of high and low intensity. Cold atoms like to settle in one or other of these regions (you can adjust which one by choosing a particular wavelength of the light). As the intensity of the light beams is turned up, the atoms fall into the 'egg-crate'-like optical traps that appear in the intensity pattern, as shown in Figure 32a. And they do so in very interesting ways.

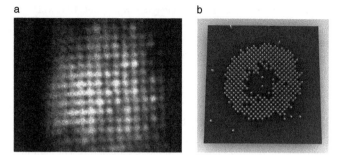

a   b

32. Cold atoms trapped in an optical lattice: a. a few hundred atoms per cell (several 10s µK), b. individual atoms at each site (a few nK).

When the atoms are cold enough, they don't like to be located at the same 'site' in this egg-crate, so the resulting distribution of atoms is very like a full egg carton—one atom at each site, as shown in Figure 32b. In this case there is no superfluidity, since the atoms like to stay put. It is more like an 'insulator', as nothing moves. By turning the light intensity up and down it is possible to explore this interesting transition between completely free flow and no flow at all.

The ability to do this in a system that is fully quantum mechanical allows scientists to explore new properties of matter that are relevant to other types of materials (for example, solid-state metal oxides) over which it is difficult to exert the same degree of precision control and measurement. In cold atomic gases it is now possible to look at atoms in these egg-crates individually and see what they are doing as changes are made to their environment.

It is possible to explore this low-temperature regime with many different kinds of atoms, and to build complicated trapping structures using light. The idea of using cold atoms to 'simulate' other quantum systems is a current area of research. It allows exploration of complex problems that cannot be solved in other ways, and is expected to lead to a new understanding of materials and structures that will have impact in new ways—perhaps helping to understand and even design new magnets that will be used for applications such as data storage for computers, magnetic resonance imaging machines for healthcare, or even friction-free motors for levitating trains.

Ultrafast

Light pulses can be extremely brief. In Chapter 5, I stated that they can be as short as a single cycle of the optical field. For light in the visible region of the spectrum, that's about 2 fs. For light in the extreme ultraviolet region, which is of shorter wavelength and higher frequency, the durations can be much shorter. The shortest yet measured is less than 100 as (10^{-18} s) long. These are currently the shortest pulses that can be controllably generated (although we can observe events that happen on a much shorter timescale by means of particle colliders). And with the advent of bursts of light in the X-ray region of the spectrum, we can expect that even shorter timescales are possible.

Since these numbers are so mind-bogglingly small, it is helpful to put them into context. The age of the universe is approximately 5×10^{17} s. Thus the ratio of one second to the age of the universe is approximately the same as the ratio of one attosecond to one second. Or to put it in an economic context, if the national debt of the US is equivalent to a second, then one cent would be equivalent to a femtosecond. On this scale, a single attosecond is virtually worthless.

What sort of things can happen on this timescale? In Chapter 4, I introduced a simple model of an atom—called the Bohr model—in which electrons 'orbit' an atomic nucleus, attracted to it by electric forces in much the same way as planets orbit the Sun, attracted by gravitational forces. The time taken to execute these orbits for simple atoms (that is, those with only a few electrons) is about 150 as. So if we want to look at this motion, we might need to use pulses shorter than this, so we don't just see a blur.

The idea of the stroboscope is the one most relevant to our story, since a variant is used by researchers today to look at the really speedy changes that go on at the fundamental level of atoms and molecules. In this application the light pulses from a laser are split into two (or more) parts, with a delay introduced between them. The first pulse in the sequence illuminates the sample, and some of it is absorbed. This 'triggers' some changes in the system—electrons move around in the atom, or bonds vibrate in molecules or solids. An instant later the second pulse arrives and some of it is again scattered from the sample and detected.

As the delay between the two pulses increases in repeated runs of this experiment, the detected scattered light maps out the dynamical changes of the sample. In a sense it makes a 'movie' of the atom or molecule or solid as it changes. This 'pump-and-probe' scheme has been used to get into the guts of what happens, for instance, during a chemical reaction, when two molecules are reconfigured by their interaction. More sophisticated versions of this kind of approach exist using several light pulses. These approaches are being used now to explore many fundamental features of extremely interesting and puzzling materials, from interacting atoms and high-temperature superconductors to biological systems.

I've noted that the shortest pulse it is possible to generate is a single cycle of the optical field. It turns out that you can devise an experiment to measure the oscillations of the optical electric field using the extreme ultraviolet (EUV) pulses produced by high-harmonic generation. What is needed to measure the pulse field is a very fast process, one that is much faster than the optical cycle itself. That can be provided by a pulse with much shorter wavelengths, about twenty or thirty times shorter than that of the optical wavelength. Pulses of such brevity are generated when an electron is ripped off an atom by means of a strong optical pulse. This requires an optical field whose strength is comparable to the binding force of the electron to the atomic nucleus. Such pulses are readily available by adding optical amplifiers to the output of a mode-locked laser.

When the electron is liberated by such an intense pulse, it finds itself sitting in a rapidly oscillating electric field, and, if its liberation occurs near a time when the field has zero amplitude, the electron can 'surf' along the next cycle of the optical wave, taking an excursion away from the atom, and then back again. When it returns it is moving very fast, and when it recollides with the atom it can be recaptured by emitting all of its extra energy as light. In this case a very short pulse is emitted as the electron recombines, having a very short wavelength of perhaps a few tens of billionths of a metre, in the EUV region of the spectrum, about twenty times shorter than the optical wave that generated it.

Now imagine that this EUV pulse shines on another atom. It has a sufficiently short wavelength such that it is absorbed by the atom, and knocks off an electron, which then sits near the atom. Imagine further that the atom is simultaneously illuminated with the short optical pulse we seek to measure. The field of this pulse accelerates the electron in one direction or another depending on the part of the optical cycle at which the electron is liberated by the EUV pulse. By changing the delay between the EUV pulse and the optical pulse, the acceleration of the electron can be measured since faster electrons, which have been accelerated to a greater degree, have more energy. In this way it is possible to 'see' an optical pulse field (Figure 33), despite the extraordinarily short timescale of the oscillations of the field.

An example of an application of the methods of pump-and-probe spectroscopy in biochemistry is the study of the first steps in the process of photosynthesis, by which plants convert carbon dioxide from the air into oxygen using sunlight as an energy source. The processes by which this happens involve transporting energy around a big biological molecule with remarkably high efficiency. The means by which this happens has some very interesting and poorly understood features—it is faster than one might expect and much more efficient. If we could learn from systems that have

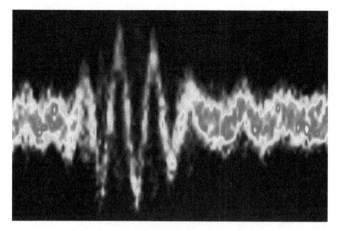

33. Direct image of the electric field of an optical pulse. The time between two adjacent peaks is approximately 2.5 fs.

evolved naturally over eons how to do this, perhaps our understanding would enable us to apply it to things like improving the design of solar cells, which would have an enormous impact on society.

Ultra-intense

Your electricity bill tells you how much energy you used in the previous month. It is measured in units of kilowatt-hours (kWh), and you are charged for each unit that you consume. Let's say that you used 220 kWh in some month (this is the average monthly energy consumption in the UK). Now, you could use all this energy at the same rate across all four weeks of the month. Or you could use it in the first week, and use nothing in the subsequent three weeks. But can you imagine using it all in a million billionth of a second? You'd need to have an awful lot of appliances to use that much energy in such a short time, and you'd need to be able to switch them on and off impossibly quickly. But the peak rate or power would be immense.

It is possible to produce light pulses that can achieve this. That is, they are incredibly brief and contain this amount of energy. In fact, it is possible to produce a pulse that delivers energy at a rate equal to the entire electricity generating capacity of the planet at a given instant. But the lights in your house won't go out, because the pulses are so brief that the total energy in them is very small.

Lasers that produce pulses of this kind are massive instruments, occupying large buildings that are a significant fraction of a football field in size. One example is the VULCAN laser at the Rutherford Appleton Laboratory in England. VULCAN produces pulses with 500 Joules of energy (3.6×10^6 J = 1 kWh) in a pulse of 500 fs duration. 500 J is the energy consumed by a 100 W light bulb in five seconds. Yet the brevity of the pulse means that the intensity of the light can be as high as a million suns. The laser at the National Ignition Facility (NIF) in Livermore, California is much, much bigger than this. And the proposed European Light Infrastructure project is set to deliver a system capable of even greater peak power than that at NIF.

Such very brief, very intense light bursts can be used to alter states of matter. The electric field at the most intense moment of the light pulse is larger than the attractive field between the electrons and nuclei that holds atoms together. So it is possible to strip the electrons off the atoms to form a new state of matter—a plasma. And it's possible to do this in an instant, shorter than the time over which the atomic nuclei can move, so that the plasma is very dense—nearly the same density as in a solid block of material such as a piece of glass, except now at two million degrees Celsius.

These are the conditions inside the cores of the giant planets and even some stars: very high-density plasmas with particles colliding with each other at high speed and at pressures of a million times that of our own atmosphere. It is possible to use this new laboratory-accessible state of matter for several things. For instance, we can begin to understand how stars work, what their life cycle is, and to characterize their stages of evolution, such as supernovae explosions and white dwarfs. Other situations of interest to astrophysicists are also amenable to experimental exploration using lasers. Astrophysicists also use such plasmas to probe the frontiers of planetary science. For instance, it is possible to infer the composition of the gas giants from their mass and size, but only if the degree to which matter can be compressed at such high pressures is known.

Some laser facilities generate pulses using light of very short wavelengths. These pulses are produced by accelerating electrons in magnetic fields, so that they 'wiggle' from side to side as they whiz down the accelerator. This produces a kind of synchrotron radiation consisting of short bursts of X-rays. Often these kinds of lasers use techniques, and indeed hardware, from particle accelerators. Examples of these are the Stanford Linear Collider Light Source (LCLS) and the Hamburg X-ray Free Electron Laser (XFEL).

The techniques afforded by the most intense laser pulses, in conjunction with short bursts of X-rays, enable scientists to diagnose plasmas in a wide range of conditions. Further, the immense pressure exerted by lasers between atomic nuclei can cause them, under the right conditions, to fuse together, releasing a large amount of energy in the process. This 'nuclear fusion' is a possible route to an almost unlimited source of energy. This application of lasers is extremely technically demanding, and is one of two methods that are being explored to achieve fusion: the other does not involve light except as a way to monitor the process. Both make use of dense plasmas.

As the laser pulses move through these plasmas, they generate a wave, very similar to the wake that trails a boat moving across a water surface. The electric fields in the plasma wave can reach more than 100,000 V over 10^{-6} m (that's about ten times the voltage in the big power lines on pylons, over a distance of less than one tenth of a human hair's breadth.) These field strengths are at least 1,000 times bigger than the accelerating fields used in the world's biggest machines for studying fundamental particles, such as the Large Hadron Collider (LHC) at CERN in Geneva. It may be possible using laser methods eventually to build table-top devices that can accelerate electrons to similar energies as currently possible in the LHC.

It is also possible to accelerate heavier particles, such as protons, by means of the extraordinarily strong electric fields generated by the interaction of laser pulses with matter. Proton beams are currently being explored as a cancer treatment, since the delivery of heavy particles to diseased tissue can be done more precisely and at greater depths than is currently possible using other types of radiation therapy.

The extraordinary properties of light continue to enable new realms of discovery, across a wide range of fields. Light is a ubiquitous tool for science and technology.

Chapter 8
Quantum light

In Chapter 1, I introduced the idea that light could be construed as a stream of particles, which I labelled 'photons' for convenience. It turns out that these are real particles, which can be produced, played with, measured, stored, and used for doing things. However, even though photons are in a sense the simplest expression of light, making individual photons is not so simple. Most light sources generate light of a different kind, for which the number of photons is not fixed.

A light bulb, for instance, produces a stream of photons that sprays everywhere. If you looked at the light going in just one direction from the bulb, and then examined just a short temporal section of the beam—a time slot, if you like—then you'd be able to count some photons in that slot. But if you repeated the experiment several times, you'd find that the number of photons was random, sometimes large and other times small. The average number of photons would be fixed, depending on the brightness of the bulb, but you'd never be able to say with certainty how many photons you would measure in the beam at a given time. That's one of the characteristics of 'classical light'—light that can be described entirely in terms of waves.

Laser light is also of this kind. The average number of photons in a pulse of laser light can be large, but for any given pulse the actual number of photons will be bigger or smaller than the average. The spread of photon numbers in a pulse is approximately the square root of the average number, so that the relative 'noise'—the variation in the number of photons in each pulse compared to the mean number over all pulses—gets smaller the higher the average number of photons.

Thus a laser beam has intrinsic intensity noise. This sets a limit on the quality of images you can get with laser illumination. The fluctuations in the laser intensity mean that detecting the separation of two points in an image is imprecise. In fact it is very imprecise for low-intensity light, where the mean photon number is small (so the object is hard to see) and the variation in photon number from frame to frame is large. The only way to get precise measurements is to look for longer, thus increasing the number of photons illuminating the object, and averaging the results over many laser pulses. The relative intensity noise is reduced by this signal averaging, leading to a better-resolved image. The precision increases in proportion to the square root of the number of photons used. This is called the 'standard quantum limit', since no classical light beam can beat it.

Quantum light, on the other hand, allows you to achieve much better results in signal averaging for the same average number of photons, since quantum light can have much lower noise than any classical light. But first, you have to build a quantum light source. There are many kinds of such a source, each producing a distinct kind of quantum light. But we might consider, to be concrete, a source that generates the primitive quantum state of light—a photon.

Single photons

So how could you make just a single, individual photon? There's a very practical scheme, invented by Otto Frisch in 1965. His idea was simple. Take a single atom and put it in its excited state (see

Chapter 5 for a discussion of how to do this). Then wait for it to drop to its ground state. When it does, it emits just one photon, since only one 'quantum' of energy can be stored in a single atom. You can tell when the atom has emitted the photon, because it recoils from the 'kick' provided by the photon's momentum. If you detect the atom moving, you can determine both that the single photon is on its way and the direction in which it is going.

Some modern quantum light sources operate in a similar way to this, only they corral the atom between two mirrors (an optical 'cavity' similar to that of a laser), and excite it very quickly so that it emits preferentially into a direction defined by the cavity axis. This makes a reliable source of single photons. It is an especially 'low noise' source, since the photons are emitted with strict regularity. If you looked at a given time slot in such a beam, you'd be able to predict with certainty how many photons would be in it—just one. Therefore the intensity is exceptionally stable—it is a 'quiet' light beam, in contrast to the 'noisy' classical one.

The idea is also used in other quantum light sources. In particular, you can construct a very simple light source using nonlinear optical effects. Specifically, there are crystals that enable one input photon with high energy to be split into two photons with lower energy, each about half of the original input photon. The probability that this fission takes place is rather small, for most materials. But since the photons are produced in pairs, you can use one as a 'herald', to signal the presence of the other (Figure 34). Such light sources are the workhorse of the field of quantum optics, which uses the quantum mechanical features of light to explore the foundations of quantum physics, as well as to enable new kinds of information technologies.

Just as classical electromagnetic waves can be polarized, so can photons. So we might find a vertically polarized (V) photon or a horizontally polarized (H) photon. These would behave just like waves, in that if we measured the polarization of the photon by

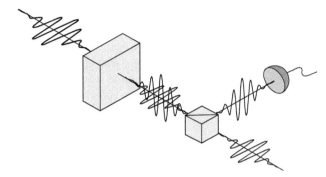

34. A 'heralded' single photon light source, generating photons randomly, but with a signal that indicates when one has been prepared.

seeing if it passed through a polarizer oriented horizontally, then we'd find that the H-photon always passed through and the V-photon never.

What's strange here is that we can construct a diagonally polarized photon, oscillating with the field at 45 degrees to both the horizontal and vertical. But if we now try to see if the photon passes through the horizontal polarizer, then there is an ambiguity. The photon is the smallest 'piece' of light, so can't be divided further. How should it behave at the polarizer? What happens is that it is transmitted with a probability of one-half, and reflected with equal probability (illustrated in Figure 35).

In practice what that implies is that if you try the experiment of putting a diagonally polarized (D) single photon into a horizontally oriented polarizer a million times, then 500,000 times it will go through. And the very strange thing about quantum mechanics is that you cannot tell on any given trial what will happen. This is not because the photon can be considered sometimes to be H-polarized and sometimes V-polarized. Rather it is because the photon is *both* H *and* V-polarized, simultaneously. The random outcomes of a measurement of the photon's polarization therefore

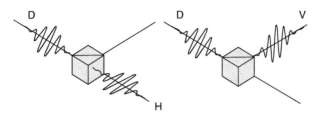

35. A diagonally polarized photon encounters a polarizer, and exits randomly through one output or the other.

reveal the intrinsic uncertainty that inhabits the most fundamental level of the universe as described by quantum physics.

Of course, you can make a virtue out of necessity in such circumstances. You can do practical things with single photons that are unimaginable with ordinary light. For instance, this property of photons can be used to generate random numbers, by measuring whether the photon is transmitted (labelling the outcome, say, 1) or reflected (labelled 0). The randomness in the string of zeroes and ones is inherent in the underlying physics, not just in the manufacture and casting of dice, or other contingencies. For this reason quantum random number generators are an emerging business—you can't fake the randomness they provide.

A second example: you can make communications links for which the security is guaranteed by the laws of physics, rather than by trusting your telecoms supplier. This is because of two important properties of photons. First, you cannot detect them in two places at once. For that reason, if an eavesdropper grabs the photon to capture the information you are sending, then of course you don't get the photon. So you receive no information, and you are aware that something's wrong.

But if the eavesdropper is clever, she will send a 'decoy' photon that she hopes will fake the message. But you can tell that it's a fake! The reason you can know this is that in quantum mechanics there is no measurement that can tell you everything about a single quantum particle.

Consider the following scenario. You want to send a simple binary message (0s and 1s) over this link, say a vertically polarized photon for 0 and a diagonally polarized photon for 1. If the eavesdropper (usually known as Eve) measures the photon and gets the answer 'vertically polarized', she could not be certain that the photon was a 0, since the diagonally polarized photon would give her the same answer at least half the time. So she gets some information, but not everything.

Now, let's say the sender of the message (conventionally called Alice—you, the receiver, are Bob) sends you a photon coded as 1. Let's say Eve measures this in the vertical orientation and gets a positive result. She must choose whether to send you a vertically or diagonally polarized photon. One strategy is to send you a vertically polarized photon, since that's the most likely source of her result. Now you measure the diagonal polarization. If your photon is from Eve, it will give you the wrong result 50 per cent of the time. If it is from Alice, you will never get the wrong result. So by comparing a section of the received message with what Alice sent, you can tell if Eve is tampering with your line.

However, Eve might be even cleverer. She may try to copy the photon from Alice without measuring it. She could make two copies in fact, sending you the original. Then she can make a vertical polarization measurement on one copy and a diagonal polarization measurement on the second, and she would have determined the full information about the photon 'bit' that Alice sent you without you ever knowing. However, she would be thwarted. A remarkable feature of quantum mechanics is that there is no possibility of building a copying machine that can do this—make an exact replica, or clone, of a single particle in an unknown quantum state. It is simply forbidden by the laws of physics. Because of these two constraints imposed by physics—'no measurement' and 'no cloning'—it is possible to build a secure communications link that can transmit a secret stream of random bits between Alice and you.

Squeezed light

There are other kinds of quantum light that provide different sorts of enhancements. Recall that light is oscillations of the electromagnetic field. A laser beam most nearly mimics this ideal behaviour. Yet even it has some 'noise' in the amplitude. That is, each time you measure the field amplitude, you get a different answer. The situation is sketched in Figure 36a, which shows some uncertainty about the field at each point, or phase, of its oscillation. There is a particular type of quantum light—called 'squeezed light'—for which this noise varies with the point in the cycle of the field, as shown in Figure 36b. It is bigger at some phases than at others. It turns out that such a field is composed of only pairs of photons. If you measure the number of photons, you will only ever find an even number. The quantum interference of all these pairs is the origin of the phase-dependent amplitude noise.

There are certain things you can do with such a state. Imagine that you wanted to make a measurement of the phase of the wave. (Recall that that is what you do in an interferometer, and the phase shift is something induced on the light beam by an object you'd like to measure, such as the presence of a particular molecule.) The phase can be determined much more precisely at points in the wave oscillation where the fluctuations of the field are smallest. In fact, the fluctuations in the squeezed light field are smaller at some phases than any classical field, so that phase sensors using such a field will be more precise than sensors using classical fields. In fact they will break the standard quantum limit.

This is a costly approach to sensing at present, so it is only used where there is a clear advantage to be had—for instance in the

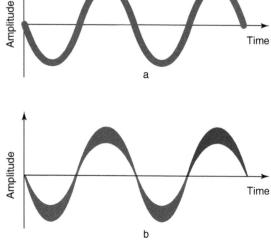

36. Squeezed light a. has reduced noise in its field amplitude at certain points in its oscillation as compared to laser light b.

detection of gravity waves by means of very large optical interferometers, such as the GEO 600 project near Hanover in Germany. By using squeezed light, this instrument can detect phase shifts that correspond to a relative path length change of the light equivalent to the size of an atom compared to the distance from the Earth to the Sun.

Quantum entanglement

Things get even stranger when we consider more than one quantum light beam. Photons can be entwined in such a way that it is impossible to ascribe a property to either of them individually—for example, a colour, position, direction, or pulse shape. This goes well beyond the fundamental notion of wave–particle duality. It challenges the very notion that in the classical world it is possible to

116

assign real values of properties to physical entities (e.g. in the case of light beams, say frequency and time of arrival, or H- and V-polarization)—in a way that can be revealed by a local measurement in a self-consistent fashion. The fact that this cannot be done for pairs of light beams prepared in certain states, and can be proven so by experiment, is one of the great triumphs of fundamental optical science in the 20th century.

By means of this property, it is possible to use quantum optics to explore the famous conjecture of Einstein, Boris Podolsky, and Nathan Rosen concerning whether a quantum mechanical description of a system of particles can be considered complete, requiring no other information to determine everything about the system. John Bell discovered in the 1960s a means to quantify such a question, and the quest to build an experiment to actually test his hypothesis began in earnest. These are known colloquially as 'Bell tests', and the earliest and currently most convincing work uses pairs of photons, each of which is correlated with the other. It is the nature of these correlations that is so different for quantum particles than for classical ones. It's worth exploring this in a bit more depth in order to get a fuller sense of the strangeness of this quantum effect.

Correlations can be found in almost every situation. Consider for instance the following simple game. A dealer takes two packs of cards, one with green backs and the other with blue backs. The dealer picks one card from each pack and gives one to you and another to your partner. Each of you looks at your card. They always have different colours on the back, of course, but they may have the same colour (red or black) on the front. In fact, you'd expect this to occur half the time, since each of you would expect individually to get either red or black with equal probability (half of the cards in each deck are black and half red).

If you and your partner found that every time you both got red or both got black, you'd say that the cards were 'correlated'. This is about as strong a correlation as you can imagine. In fact, if you both got the same colour more than half the time, you'd also be able to claim the cards were correlated, though clearly the correlations would be 'weaker' than in the first instance. By measuring the correlations, you could determine whether the dealer was cheating, since you might assume she'd start with two independent, complete decks.

We can make an analogy of this kind of correlation for photons using polarization instead of suit for the cards. That is, a horizontally polarized photon might be termed a 'red' photon, and a vertically polarized photon a 'black' one. Then if a source produces photons two at a time, as described above, you can say that it produces correlated photon beams if it always produces photons with a prescribed polarization, say one vertical and one horizontal, or both horizontal. This type of correlation is termed 'classical', since it has a complete analogy to the situation with classical objects like playing cards.

There is a feature of correlations that has an intrinsic quantum mechanical aspect. Let's say there are two possible states in which the photon pair can be prepared—the first H-polarized and the second V-polarized or vice versa. In the classical world these two situations for two particles are mutually exclusive: either HV *or* VH is possible, each with a probability of one-half. But, just as the single photon can be in a superposition H *and* V, so can the pair: HV *and* VH, shown in Figure 37. This turns out to be a much stronger correlation than is possible with any classical particles, and is called entanglement. It is the most enigmatic property of quantum physics, and has extraordinary consequences.

These are revealed by means of Bell tests. In such a test, you have to consider not only the possibility of correlations in the H and V polarizations of each particle, but also those in the diagonal (D) and anti-diagonal (A) polarizations, each oriented half way

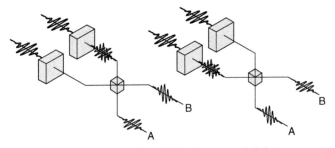

37. A light source for generating polarization entangled photons.

between the horizontal and the vertical. (Diagonally polarized
light, for example, is shown in Figure 35. Anti-diagonal polarization
is oriented at right angles to the D direction.) The analogy with
the cards is that you can look at the front of the cards and observe
either red (equivalent to H) or black (equivalent to V) suits. Or
you could look at the backs and see green (equivalent to D) or blue
(equivalent to A).

A quantum game

Now imagine a card game in which the dealer chooses from either
pack and gives one card to each player. That means that each
player will have a card that could be either red (R) or black (B) on
the front (F) and either green (g) or blue (b) on the back (B). The
dealer chooses to hand out cards in such a way that if one player
looks at the front of his card, and the other the back of hers (F,B),
then they never find the result (R,b). Similarly if the first player
looks at the back of his card and the other the front of hers (B,F),
then they never find the outcome (b,R). However, when they both
look at the front of their cards (F,F) they sometimes see (R,R).
From this, you would conclude logically that in such a case, had
they looked at the front of their cards (B,B) they would have seen
(g,g). That's what would happen for obviously classical things
like cards.

Measurement choice:	Outcomes: Player 1; Player 2; Probability			
F, F	R; R; 1/2	R; B; 1/12	B; R; 1/12	B; B; 3/4
F, B	R; b; 0	R; g; 1/6	B; b; 2/3	B; g; 1/6
B, F	b; R; 0	b; B; 2/3	g; R; 1/6	g; B; 1/6
B, B	b; b; 1/3	b; g; 1/3	g; b; 1/3	g; g; 0

38. Table of the probabilities of possible outcomes for a quantum card game.

But in fact, when you take photons (or other particles) that are quantum correlated and do such an experiment it doesn't turn out that way. What happens is that when the players make measurements of the polarizations using, for the first player, a horizontally oriented polarizer, and, for the second player, a diagonally polarized photon (or vice versa), they never get the results (V = 0, D = 1) and (D = 1, V = 0). Likewise, when they both measure using diagonally oriented polarizers, they sometimes get the result (D = 1, D = 1). Therefore you would logically conclude that when they measured the photons using a horizontally oriented polarizer, they would sometimes get the result (H = 1, H = 1). But, when they do this experiment, they never get this outcome! The table of possible results of such a quantum card game are shown in Figure 38. Such experiments can, and have been, done using photon pairs. It's not fiction.

Local properties of things

So what is going on? This is the fundamentally weird thing about quantum physics: the conclusion of the quantum card game is that the photons cannot have predetermined values of their polarization when they are prepared at the source. It is as if the cards could not have been of definite suits from a deck with a well-specified card-back colour. This goes against all intuition about cards: they surely have definite properties of a specific suit on the front of each card and a specific colour on the back. No matter whether we know or even the dealer knows what these values are or not, we don't doubt that the cards actually have these properties when they are given to us. And we certainly don't expect that anything we do to them changes those properties. But quantum mechanics tells us that we cannot assign colours to the cards a priori.

It is the measurements that give definiteness in the outcomes. We cannot claim the measurements simply reveal predetermined properties of the photons, which are unknown to the players. It is actually that you cannot assign definite polarizations to the individual photons when they are produced by the source in such a way as to give the outcomes that are actually seen. If you try to devise a way of dealing cards that gives such a result, you'll find that it is impossible. The cards would need to have the possibility that they can be simultaneously in superpositions of red and black or green and blue, in particular ways. Just so the photons—it is necessary for them to be in superpositions of H and V in a way that gives very specific types of correlations. It is this type of correlation that is termed 'quantum entanglement'.

Entanglement is a very weird concept. It is not possible to find a way to think about it in terms of common everyday objects—as the playing card example was intended to show. Yet entanglement is also very common. It appears in many things at the quantum scale, even in everyday conditions: in the correlations between electrons in molecules, giving rise to bonds between the atoms making up the molecule, or even relatively small atoms themselves, as well as exotic materials like superconductors.

Surprisingly, entanglement also turns out to have technological implications. You'd hardly think that such an arcane and abstract idea could possibly have any application, but it does. It enables a raft of information processing approaches that cannot be replicated by sending classical waves back and forth.

Indeed, the very idea that all information processing systems are at bottom built of something suggests that the design principles of these machines must reflect the underlying physics of the constituent parts—usually classical physics. This has led to the understanding that basing computing, communications, and measurement on quantum mechanics provides new opportunities for technologies that surpass those of the current generation in unimaginable ways: communications whose security is guaranteed by the laws of nature; computers that can solve 'uncomputable' problems; imaging systems that reveal an object that they are not even looking at.

Light plays an important part in implementing such systems. The infrastructure of optical fibre networks, for instance, can be used to distribute random 'quantum keys' (random strings of 0s and 1s) completely securely between two parties, which can then be used to encode messages. Such networks can also be used to connect small-scale quantum processors, eventually becoming a distributed quantum computer. Indeed, it has been shown that in principle it is possible to build a quantum computer completely out of light, though it is extremely challenging to do so. Combining these technologies holds the promise in the future of a quantum Internet, a radically different way to communicate and process information from the technology we currently use, and all enabled by light.

Chapter 9
Twilight

The atoms of Democritus
And Newton's particles of light
Are sands upon the Red Sea shore
Where Israel's tents do shine so bright.

William Blake, 'Mock on, Mock on,
Voltaire, Rousseau'

It is inevitable that in a book of this brevity—covering a topic with as long a history and as broad a range of applications as light—many things simply can't be covered. In particular, the fantastic discoveries arising from looking at other regions of the electromagnetic spectrum than the visible, and the ubiquity of optical devices in everyday life are very incompletely described.

The myriad versions of imaging devices, from multi-mirror telescopes (with diameters of tens of metres, and adaptively controlled reflectors to null the effects of the twinkling sky) to massive synchrotrons (accelerating electrons to the point that they radiate intense X-rays for looking at tiny material structures, both man-made and natural, to reveal the structure of, for instance, biologically important molecules or engineered metals) are only hinted at. But no matter: I hope that you are convinced of the beauty of light, and that a tour through its mysteries and how they were unravelled is as interesting a story as you could want.

Meanwhile, the science and technology of optics is vibrant, opening up new areas for exploration and new applications, often with unexpected fruitfulness. Light continues to reveal new mysteries and to inspire new devices: it still engages and fires our imagination as it has done for centuries.

Further reading

Non-technical books

O. Darrigol, *A History of Optics from Greek Antiquity to the Nineteenth Century*, 2012 (Oxford: Oxford University Press).

H. E. Edgerton and J. R. Killian, *Moments of Vision: The Stroboscopic Revolution in Photography*, 1979 (Cambridge, MA: The MIT Press).

J. P. Harbison and R. E. Nahory, *Lasers: Harnessing the Atom's Light*, 1997 (New York: Scientific American Library, W. H. Freeman & Co).

J. Hecht, *City of Light: The Story of Fiber Optics*, 2004 (New York: Oxford University Press).

J. Hecht, *Beam: The Race to Make the Laser*, 2005 (Oxford: Oxford University Press).

M. Kemp, *The Science of Art: Optical Themes in Western Art from Brunelleschi to Seurat*, 1992 (New Haven: Yale University Press).

A. Zajonc, *Catching the Light*, 1993 (Oxford: Oxford University Press).

Specialist texts

M. Bass et al., *Handbook of Optics*, 2000 (New York: McGraw-Hill).

M. Born and E. Wolf, *Principles of Optics*, 7th ed., 1999 (Cambridge: Cambridge University Press).

R. W. Boyd, *Nonlinear Optics*, 3rd ed., 2004 (New York: Academic Press).

J. W. Goodman, *Statistical Optics*, 2000 (New York: Wiley Interscience).

J. W. Goodman, *Introduction to Fourier Optics*, 2004 (New York: Roberts & Co.).

H. Hariharan, *Basics of Interferometry*, 2nd ed., 2003 (New York: Academic Press).

S. Haroche and J.-M. Raimond, *Exploring the Quantum: Atoms, Cavities and Photons*, 2006 (Oxford: Oxford University Press).

E. Hecht, *Optics*, 4th ed., 2001 (Reading, MA: Addison Wesley).

S. M. Hooker and C. Webb, *Laser Physics*, 2010 (Oxford: Oxford University Press).

J. Mertz, *Introduction to Optical Microscopy*, 2010 (New York: Roberts & Co.).

L. Novotny and B. Hecht, *Principles of Nano-optics*, 2nd ed., 2012 (Cambridge: Cambridge University Press).

V. Vedral, *Introduction to Quantum Information Science*, 2006 (Oxford: Oxford University Press).

W. Welford, *Optics*, 3rd ed., 1988 (Oxford: Oxford University Press).

Light

Index

Light

Light

Index

SOCIAL MEDIA
Very Short Introduction

Join our community
www.oup.com/vsi

- Join us online at the official Very Short Introductions **Facebook** page.
- Access the thoughts and musings of our authors with our online **blog**.
- Sign up for our monthly **e-newsletter** to receive information on all new titles publishing that month.
- Browse the full range of Very Short Introductions online.
- Read **extracts** from the Introductions for free.
- Visit our library of **Reading Guides**. These guides, written by our expert authors will help you to question again, why you think what you think.
- If you are a teacher or lecturer you can order inspection copies quickly and simply via our website.